AF845415

Current Trends in Short- and Long-period Fiber Gratings

Current Trends in Short- and Long-period Fiber Gratings

Editor

Edvin Falk

scitus academics

Current Trends in Short- and Long-period Fiber Gratings
Edited by **Edvin Falk**

ISBN: 978-1-68117-288-0
Library of Congress Control Number: 2016937464

© 2017 by
SCITUS Academics LLC,
www.scitusacademics.com
Box No. 4766, 616 Corporate Way,
Suite 2, Valley Cottage,
NY 10989

This book contains information obtained from highly regarded resources. Copyright for individual articles remains with the authors as indicated. All chapters are distributed under the terms of the Creative Commons Attribution License, which permits unrestricted use, distribution, and reproduction in any medium, provided the original author and source are credited.

Notice

Reasonable efforts have been made to publish reliable data and views articulated in the chapters are those of the individual contributors, and not necessarily those of the editors or publishers. Editors or publishers are not responsible for the accuracy of the information in the published chapters or consequences of their use. The publisher believes no responsibility for any damage or grievance to the persons or property arising out of the use of any materials, instructions, methods or thoughts in the book. The editors and the publisher have attempted to trace the copyright holders of all material reproduced in this publication and apologize to copyright holders if permission has not been obtained. If any copyright holder has not been acknowledged, please write to us so we may rectify.

Printed in United States of America on Acid Free Paper

Preface

A fiber Bragg grating (FBG) is a type of distributed Bragg reflector built in a short segment of optical fiber that reflects specific light wavelengths and transmits all the others. This is achieved by creating a periodic variation in the fiber core refractive index, which generates a wavelength-specific dielectric mirror. A Bragg fiber grating can therefore be used as an inline optical filter to block certain wavelengths or as a wavelength-specific reflector. Fiber Bragg gratings are created by the systematic "inscription" or "writing" of the refractive index into the core of a special type of optical fiber using an intense ultraviolet (UV) source such as a UV laser. A long-term fiber grinding pair illuminates from a guided mode to propagate cladding modes where absorption and dispersion cause it to be lost. It is an optical fiber structure with properties that vary periodically along the fiber so that several propagating modes meet the interaction conditions. This structure's time is in the order of a millimeter fraction. Unlike Bragg fiber gratings, LPFGs copropagate pairs with close propagation constants, which mean that the time of such gratings can exceed the wavelength of the fiber radiation significantly. As the LPFG period is much longer than the wavelength, the LPFGs can be produced relatively easily. Since LPFGs copropagate multiple modes, their resonance can only be seen in the transmission spectrum. Long-term grating has a wide range of applications, including filters, flattening filters and sensors for tape repelling. Various gratings have been designed with complex structures: gratings that combine multiple LPFGs, LPFGs with superstructures, chirped gratings and apodized grates. Various LPFG-based devices have been developed: filters, sensors, fiber dispersion compensators, etc.

This book highlights current trends and research in both short and long fiber gratings. This book aims to highlight state of the art topics in this field and shows the recent developments in fiber gratings.

Table of Contents

CHAPTER 1

Surface-Corrugated Microfiber Bragg Grating

Fei Xu[1], Jun-Long Kou[1], Yan-Qing Lu[1], Ming Ding[2] and Gilberto Brambilla[2]

[1] College of Engineering and Applied Sciences and National Laboratory of Solid State Microstructures, Nanjing University, Nanjing, P. R. China

[2] Optoelectronics Research Centre, University of Southampton, Southampton, SO17 1BJ, United Kingdom

1. INTRODUCTION

Optical fiber Bragg gratings (FBG) are key devices for optical communication and sensors. FBGs are based on a periodic variation in the refractive index of the fiber core and allow for the reflection of a narrowband signal centered at a specific Bragg wavelength. Over the last two decades, FBGs have been manufactured mainly by modifying the core refractive index using interferometric or point-by-point techniques; most of interferometric techniques use a phase mask and an ultraviolet (UV) laser [1] (typically excimer or frequency doubled Ar+ ion) or femtosecond (fs) lasers (near IR [2] or UV [3]). Gratings based on surface corrugations have also been demonstrated in etched fibers using photolithographic techniques [4]. All gratings fabricated in thick fibers have weak refractive index modulations (Δn_{mod}~10^{-4}-10^{-3}) and the related grating lengths are of the order of several millimeters.

Structural miniaturization is one of the current trends for achieving higher-bandwidth, faster response and higher-sensitivity. Experimentally, FBG miniaturization has been achieved in two steps. Firstly, the fiber has been tapered into a subwavelength-scale microfiber (MF), considered to be the basic element for miniature fiberized devices and sub-systems [5]. Then strong refractive index modulations ($\Delta n_{mod}>10^{-1}$) are induced. Large Δn_{mod} can be obtained by alternating layers of different materials, one of which can be air. Although this process in normal optical fibers imposes the removal of large amounts of material (the propagating mode is confined at a depth >50 μm from the fiber surface), in fiber tapers and tips it only requires the removal of ~10,000 less matter because the propagating mode is confined by the silica/air interface in areas with micron size. A few techniques have been proposed to fabricate surface-corrugated fiber gratings (SCMGs), including photorefractive inscription using etching [6], femtosecond lasers [7, 8], and focused ion beam (FIB) [9-16]. So far, FIB is the most flexible and powerful tool for patterning, cross-sectioning or functionalizing a subwavelength circular microfiber due to its small and controllable spot size and high beam current density. In the past two years, a number of ultra-compact SCMGs have been successfully fabricated by FIB milling, with lengths as small as ten micrometers. In addition, a variety of other techniques have been proposed to generate gratings exploiting the fraction of power propagating in the evanescent field: these include wrapping a microfiber on a microstructured rod or put a microfiber on a surface-corrugated planar grating. This chapter reviews recent advances in ultra-small SCMGs, their characteristics and applications.

2. THEORY

2.1. SCMG spectral properties

In SCMGs, strong Δn_{mod} is achieved alternating layers of air and glass. Figure 1 shows a typical SCMG geometry: the surface-corrugated structure is fabricated by inducing periodic microscale open-notches on the side or holes across the optical microfiber.

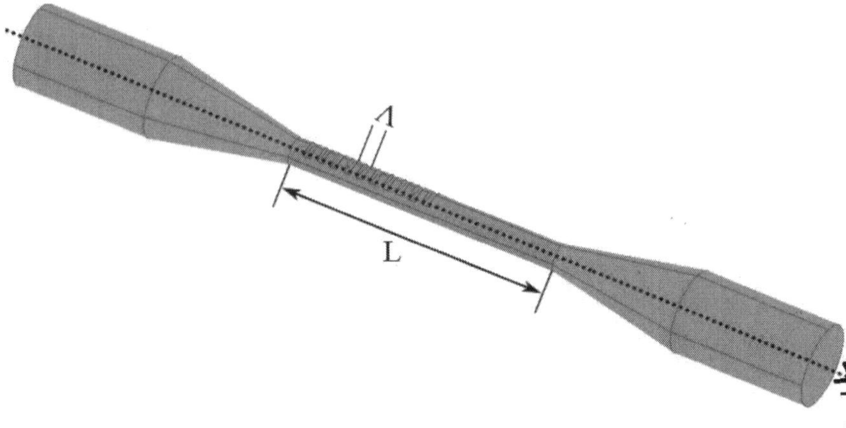

Figure 1. Schematic illustration of SCMGs. L and Λ are the grating length and period, respectively. $\vec{j}$ is the unit vector along the fibre longitudinal axis.

In the grating, the forward (f) and backward (b) propagating modes are related by

$$\overrightarrow{\beta_b} = \overrightarrow{\beta_f} + m\frac{2\pi}{\Lambda}\vec{j}. \tag{1}$$

where $\beta_i = (2\pi/\lambda)n_{eff,i}$ (i = f or b) is the mode propagation constant, m is the diffraction order, Λ is the period of the grating and $\vec{j}$ is the unit vector along the fibre longitudinal axis. In a more physical perspective, Eq. (1) means that the momentum mismatch between the forward and backward propagating modes should be compensated by the reciprocal vector provided by the periodical index modulation. For the first-order diffraction which is commonly seen in SCMGs:

$$\left(n_{eff,f} + n_{eff,b}\right)\Lambda = \lambda_B. \tag{2}$$

Furthermore, if the two modes are identical, the commonly used Bragg resonance condition can be obtained, namely, $\lambda_B = 2n_{eff}\Lambda$.

For a SCMG, the effective index difference between the MF milled and unmilled cross sections can be as large as ~ 10^{-3}[9], or even ~ 10^{-1}[12], orders of magnitude larger than that in conventional FBGs. One way to calculate an averaged effective index of the grating region is to choose an unperturbed waveguide boundary using the method developed by W. Streifer [17]: the boundary between different materials is shifted to compensate for different geometry, shown as the dashed line in Fig. 2. d is the depth of the corrugation and h_{eff} is the boundary shift from the top of the corrugation to the new boundary of the corresponding unperturbed waveguide. The boundary shift h_{eff} illustrated in Fig. 2(b) and (c) is determined so that the volume bounded by the upper part of the corrugation ($S_A\tau$) is equal to the volume bounded below [$S_B(1-\tau)$], *i.e.*

$$S_A \tau = S_B (1-\tau). \tag{3}$$

where τ is the duty cycle. The averaged effective index could thus be obtained by mode analysis after the unperturbed waveguide boundary is established.

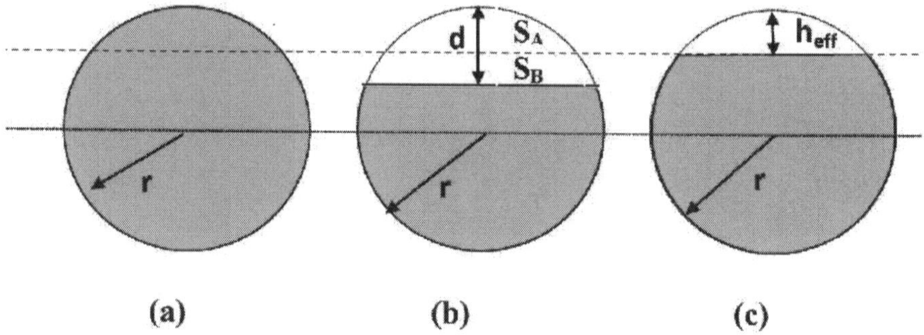

Figure 2. Equivalent unperturbed geometry. The cross-sections of (a) an unmilled MF, (b) a milled MF and (c) an equivalent unperturbed geometry, respectively. d is the groove height and h_{eff} is effective height, obtained solving eq. 3..

The SCMG reflection spectrum can be estimated using [18]. However, due to the large index difference in the corrugation region, strong scattering may occur. A more effective way to verify the experimental spectrum obtained from the SCMG is a 3D finite element simulation, as shown in Fig. 3. This method takes the details of the structural deformation into consideration and thus better reflects the real situation experienced by the light field.

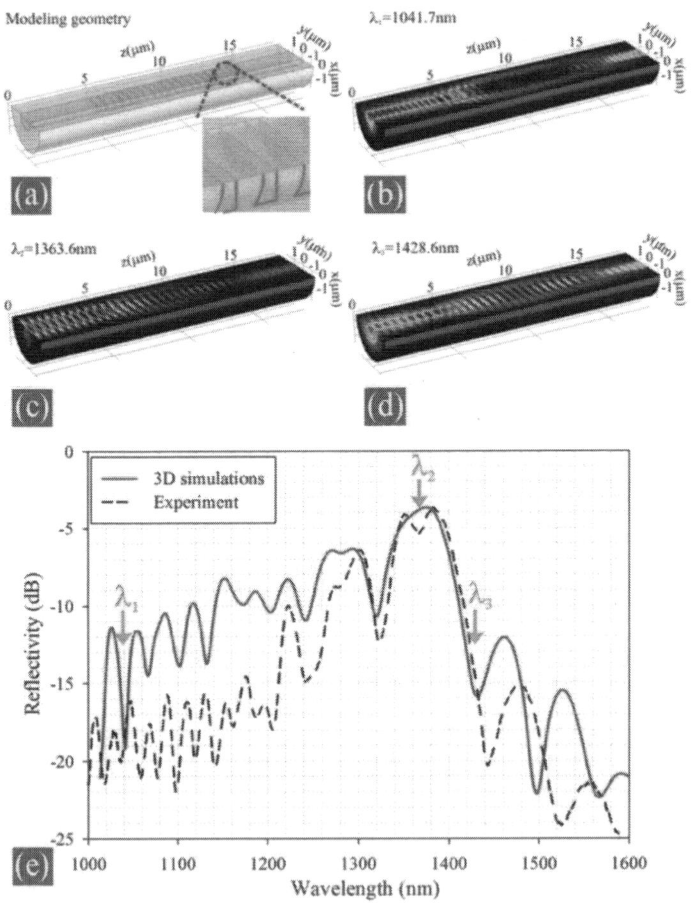

Figure 3. a) Schematic of a 3D finite element simulation of SCMG; the insert shows the magnified figure of the biconcave air notch. (b), (c), and (d) Electric fields at wavelength λ_1 = 1041.7nm, λ_2 = 1363.6nm, and λ_3 = 1428.6nm, respectively. (e) SCMG reflection spectra. The red solid line is the 3D simulation line while the blue dashed line is the experiment result. λ_1, λ_2, and λ_3 represent the wavelengths whose electric fields are shown in (b), (c), and (d) [*19*].

2.2. SCMG sensing mechanisms

In the past years, FBGs have been widely applied to the measurement of chemical, biomedical, physical, electrical parameters, especially for structural health monitoring in civil infrastructures, where the information of measurands usually rely on the monitoring of the shift in wavelength of the reflected Bragg signal with the changes in the measurand (*e.g.*, refractive index, temperature, and strain/force). FBG based sensors have a number of advantages with respect to conventional sensors, such as compactness, immunity to electromagnetic interference, multiplexing, rapid response to real time monitoring, and high sensitivity to external perturbations. Compared with conventional FBGs, SCMGs have similar applications and advantages with some extra benefits such as ultra-compact size and large evanescent field.

As the effective index and period of grating is a function of r, n_a, T, and ε, the Bragg condition (eq. 2) can be rewritten as [19]

$$\lambda_B = 2n_{eff}(r, n_f, n_a, T, \varepsilon)\Lambda(T, \varepsilon). \tag{4}$$

where the refractive index of fused silica and of the ambient medium surrounding the SCMGs are denoted by n_f and n_a, T is the temperature and ε is the strain applied to the SCMGs.

2.2.1. Refractive index

The notable distinction between SCMGs and conventional FBGs lies in the large SCMG evanescent field which enables its capabilities for external medium sensing. When a SCMG is operated as a refractive index (RI) sensor, the wavelength shift depends on the change of n_a. The sensor sensitivity (S_a) with respect to the ambient medium RI is defined as [19]

$$S_a = \frac{d\lambda_B}{dn_a} = \frac{\partial\lambda_B}{\partial n_{eff}(n_a, r_{MF})}\frac{\partial n_{eff}(r_{MF}, n_a)}{\partial n_a} = 2\Lambda\frac{\partial n_{eff}}{\partial n_a}. \tag{5}$$

Figure 4 shows S_a as a function of r_{MF}. n_a is chosen to be 1.33 and 1.42, because most of the RI sensors work around these values. S_a increases for decreasing MF diameters, as larger fractions of power propagate in the evanescent field, thus in the surrounding environment. For the same reason, higher external RIs are associated with a larger S_a. Theoretically, the largest S_a which can be obtained by SCMG is ~2Λ, typically around 1100 nm/RIU according to Eq. (5). This value is comparable to optical microfiber coil resonator sensor [20, 21] and higher than sensors based on microcapillary resonator [22] or photonic crystal microresonator [23].

2.2.2. Temperature

Temperature affects the Bragg wavelength shift through the thermo-optical effect and thermal expansion in three ways: index variation, MF radius variation and grating period change, each of which is represented in Eq. (6). The temperature sensitivity (S_T) can be defined as [19]:

$$S_T = \frac{d\lambda_B}{dT} = 2\Lambda\left(\sigma_T \frac{\partial n_{eff}}{\partial n_f} + r\alpha_T \frac{\partial n_{eff}}{\partial r} + n_{eff}\alpha_T\right) \tag{6}$$

Here, σ_T (1.2×10^{-5}/°C) is the thermo-optical coefficient and α_T (5.5×10^{-7}/°C) is the thermal expansion coefficient of fused silica. As thermal expansion contributes less than 2 pm/°C to the total sensitivity, it is generally neglected. S_T resulting from the thermo-optical effect is ~ 10 – 20 pm/°C and dominates in temperature sensing, which is in agreement with previous results obtained using fiber tip Fabry-Perot interferometers [24].

2.2.3. Strain/Force

From continuum mechanics, when longitudinal strain is applied to a SCMG, its Bragg wavelength shift can be estimated as follows [19]:

$$\Delta\lambda_B = 2\Lambda n_{eff}\left\{1-\left(\frac{n_{eff}^2}{2}\right)\left[p_{12}-\nu(p_{11}+p_{12})\right]\right\}\varepsilon = 2\Lambda n_{eff}\left(1-p_{eff}\right)\varepsilon, \tag{7}$$

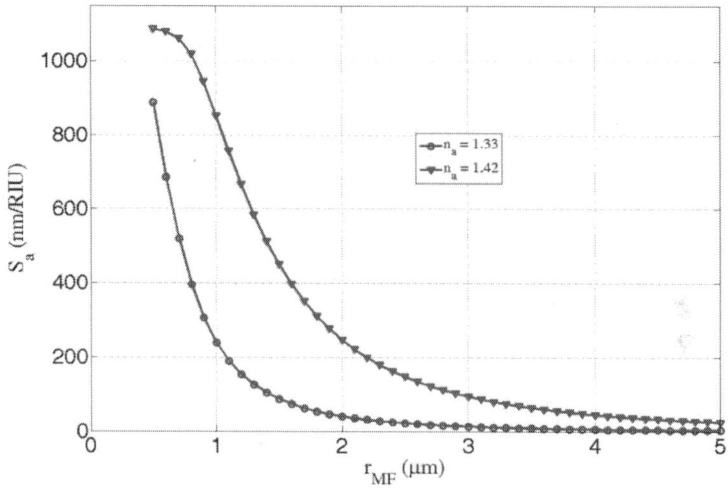

Figure 4. Dependence of sensor sensitivity S_a on the MF radius r_{MF} in the range 0.5 – 5μm for two values of surrounding medium refractive indices n_a. Resonant wavelength is set at 1550nm and n_a is chosen to be at 1.33 (blue circles) and 1.42 (red triangles). Only the fundamental mode is considered.

where ε is the applied strain, *v* is Poisson's ratio and p_{ij} coefficients are the Pockels' strain-optical tensor coefficients of the fiber material. If the MF cross section deformation due to the applied strain is neglected, strain sensitivity (S_S) is reduced to

$$S_S = \frac{\Delta\lambda_B}{\varepsilon} = \lambda_B\left(1-p_{eff}\right). \tag{8}$$

The effective photo-elastic coefficient p_{eff} for a SCMG strain sensor is ~ 0.21, giving S_S ~ 1.2 pm/με, which is comparable to that of a conventional fiber grating with a Bragg wavelength of 1550 nm. This is in agreement with experimental results [25, 26]. SCMG sensors can also be characterized by their force sensitivity (S_F)

$$S_F = \frac{S_S}{\pi r_{MF}^2 E}, \tag{9}$$

where E represents the fibre Young's modulus. Equation (9) shows that S_F scales inversely with the square of the microfiber diameter.

For composite structures, like the SCMG wrapped on a PCF mentioned below, sensitivity depends on more parameters such as the strain-optical tensor of the photonic crystal fiber (PCF) support and the embedding coating.

3. SCMG FABRICATION

Several techniques have been reported in the literature for the fabrication of gratings in MFs and they can be classified as follows [19]:

1. Etch-eroded commercial FBG or UV irradiated FBG [25-33].
2. FIB-milled FBG on MFs [9-16].
3. Femtosecond-laser-irradiated FBG on MFs [7, 8].
4. FBGs exploiting evanescent fields [34-37].

Both cladding-etched commercial FBGs and UV irradiated FBGs in MFs are uniform microfiber gratings, meaning that the grating region experiences only RI modulation and not structural perturbations as indicated in Fig. 1. The most commonly used technique to get an uniform microfiber grating is to etch a single mode fiber (SMF) after the FBG has been written in the photosensitive Ge-doped core [25, 27-29]. Usually, a hydrofluoric acid aqueous solution (~ 20% – 50%) at room temperature is employed for the etching process at an etching speed of ~ 0.5 – 2 μm/min. The diameter of the etched fiber can be measured and controlled in situ by monitoring the transmission loss.

Femtosecond-laser-irradiation is another way to induce periodical physical deformations on the surface of MFs [8]. During the femtosecond laser irradiation, the ultra-short laser pulses transfer energy to the electrons in the material irradiated through nonlinear ionization [38]. When a sufficiently high energy is achieved, pressure or shock waves cause melting or non-thermal ionic motion, resulting in permanent structural damages in the material. Aided with proper phase masks, uniform microfiber gratings can be fabricated on the surface of the MFs [7, 8].

For the fabrication of ultra-small SCMGs, the main technique is FIB milling or inducing external gratings.

3.1. FIB-milled SCMGs

FIB milling, a powerful micromachining technique, has been the tool of choice to fabricate SCMGs [9-16]. This method employs accelerated ions to mill nanometer-scale features on MF surfaces to form corrugated structures. As the index modulation results from changes in the structure, this kind of gratings are called structural SCMGs.

Prior to milling, the MF is coated with a thin film of metal, *e.g.* aluminum or gold [9-13], to prevent charge accumulation which cause ion deflections and large fabrication errors. Alternatively, MF can also be laid on a doped silicon wafer [14]: due to van der Waals' forces, the MF tightly attaches to the conductive substrate and it avoids charging by transferring charges to the wafer.

During the FIB micromachining process, the MF sample should be fixed firmly in the vacuum chamber to minimize sample displacements. A 30kV, 10 – 300pA Ga^+ ion beam is usually used to get a good milling accuracy. The total milling process takes minutes to hours according to the beam current used and milled area. After the machining process, when metal coating is not required, the MF is immersed in metal etchant to totally remove the metal film and then is cleaned with deionized water.

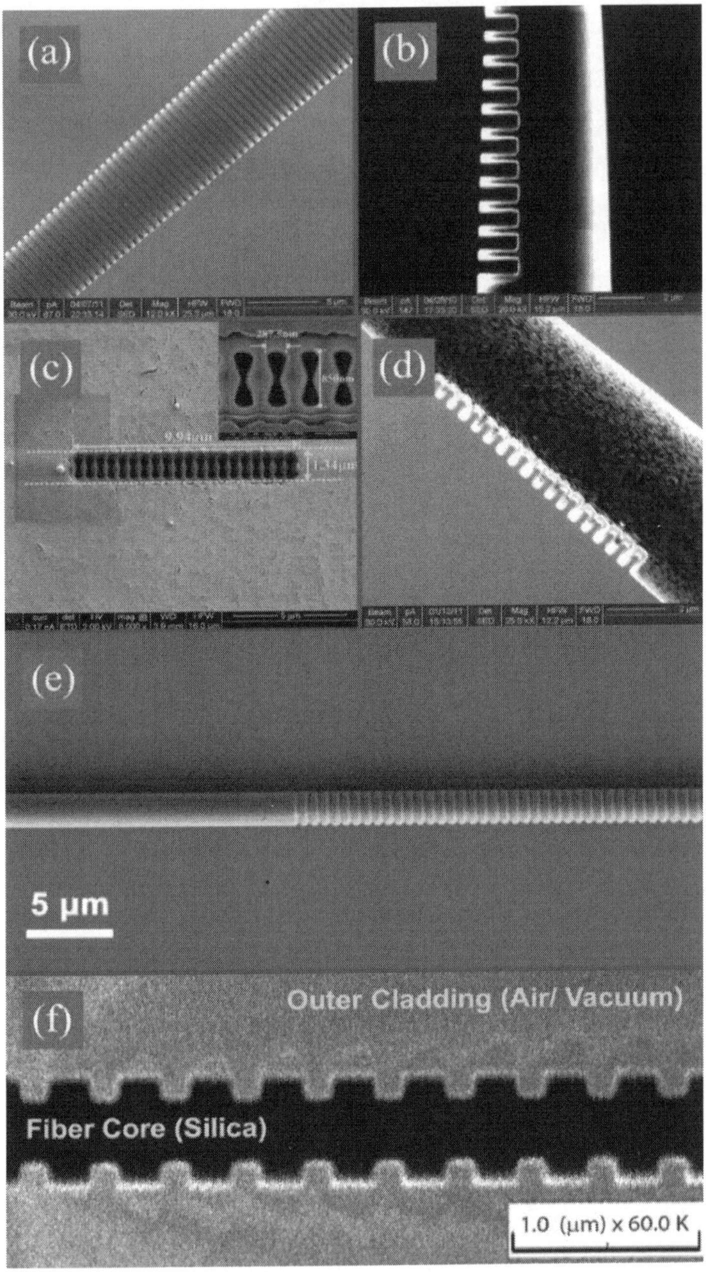

Figure 5. FIB/SEM pictures of gratings fabricated on MF tips (a)[9], (b)[10], (d)[13] and MF tapers (c)[12], (e)[14], (f)[15]. Reprinted with permission. Copyright 2011 Optical Society of America, Copyright 2012 AIP and Copyright 2011 IEEE.

Figure 5 shows FIB/SEM pictures of SCMGs fabricated from different groups. Gratings in Fig. 5(a), (b), and (d) are fabricated on MF tips while the rest are on tapers. SCMGs are fabricated on MFs with diameter ranging from 560 nm [Fig. 5(f)] to 6.6 μm [Fig. 5(a)] and the number of grating periods varies from 11 [Fig. 5(d)] to 900 [Fig. 5(e)]. Both high [5×10^{-3} – 10^{-1}, Fig. 5(c), (d), (f)] and low [10^{-4} – 5×10^{-3}, Fig. 5(a), (b), (e)] average RI modulations have been achieved by FIB-milled SCMGs. In all, FIB provides researchers a flexible way to get all kinds of structures with high accuracy at will and without additional masks. Yet, batch production cannot be envisaged for this method.

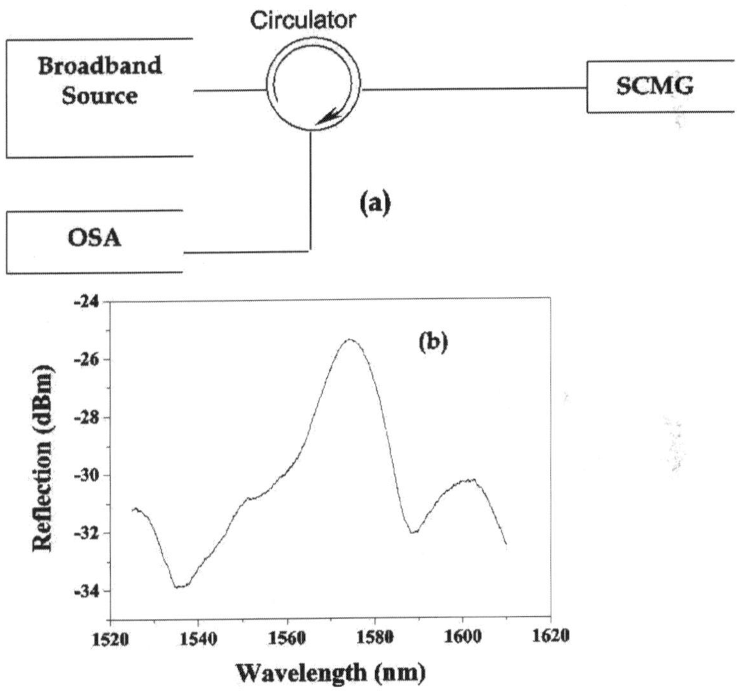

Figure 6. a) Experimental set-up of characterizing SCMGs. (b) Reflection spectra of a SCMG in air.

Experimentally, SCMGs are characterized with a set-up similar to that shown in Fig. 6(a): light from a broadband source is injected into the grating and the reflected light is then collected by an optical spectrum analyzer (OSA) (Ando AQ6317B, Japan) after passing through a circulator. The fiber tip reflection has been analyzed before grating inscription and displayed a negligible reflection over the whole spectrum, showing that the reflection at the tip end was insignificant. Fig. 6(b) shows the typical reflection spectrum of a SCMG. There is a clear peak at ~ 1572 nm with an extinction of 10 dB. Two side lobes are observed at longer (1600 nm) and shorter (1535 nm) wavelengths. The possible explanation of the wide bandwidth of the experiment reflection spectrum can be the effective indices variation, rough notch surfaces and departure from perfect grating periodicity.

3.2. SCMG exploiting evanescent fields

In addition to the previous techniques, other methods have demonstrated or proposed the fabrication SCMGs using the strong microfiber evanescent field. SCMGs can be manufactured by wrapping a MF on a microstructured rod and then coating them with a low-loss polymer. The rod can be obtained by etching a cane used in the manufacture of microstructured fibers or it can be a thick microstructured fiber. Here no expensive mask, laser or machining set-up is used. Combining current enabling technologies on microstructured optical fibers [39, 40] and MFs [41], it is possible to induce periodic or variable corrugations leading to the coupling between forward and backward propagating waves. The final SCMG structure is shown in Fig. 7(a) [34, 37]: it is a compact and strong micro-device with some air holes arranged in a circle. Fig. 7(b) presents a possible rod cross-section. The mode propagating in the microfiber experiences refractive index corrugations because its evanescent field overlaps with the support rod where there is alternation of glass and air holes. Here we only consider one ring of air holes; the inner large hole is also filled with air and the MF is assumed perpendicular to the rod. When a coordinate increasing along the microfiber is used, the surface corrugations experienced by the mode propagating along the curvilinear coordinate are similar to those experienced by a mode propagating straight in proximity of a conventional planar grating. Unfolding the MF, the MF grating can be taken as a coated MF on a planar substrate with air-hole corrugations: the equivalent structure is shown in Fig. 7(c) [34, 37].

The mode field in the unperturbed waveguide geometry is derived from the perturbed (straight) geometry using the method mentioned in section 2.1. If g_1 is the boundary between the materials with refractive indices n_1 (air) and n_2 (the rod material, generally silica) and g_2 is the boundary between the materials with refractive indices n_4 (coating) and n_2, a new g_0 is chosen to match the volume of n_1 material extending into region A (above g_0) with the volume of n_2 material extending into region B (below g_0), as shown in Fig. 8(a). We introduce the effective distance d_{eff} between g_2 and g_0 as[34, 37]:

$$d_{eff} = g_1 + g_2 - g_0 = d_1 + d_2 + 2r - \pi r^2 / \Lambda \tag{10}$$

d_{eff} increases for increasing $d = d_1+d_2$ and increasing Λ.

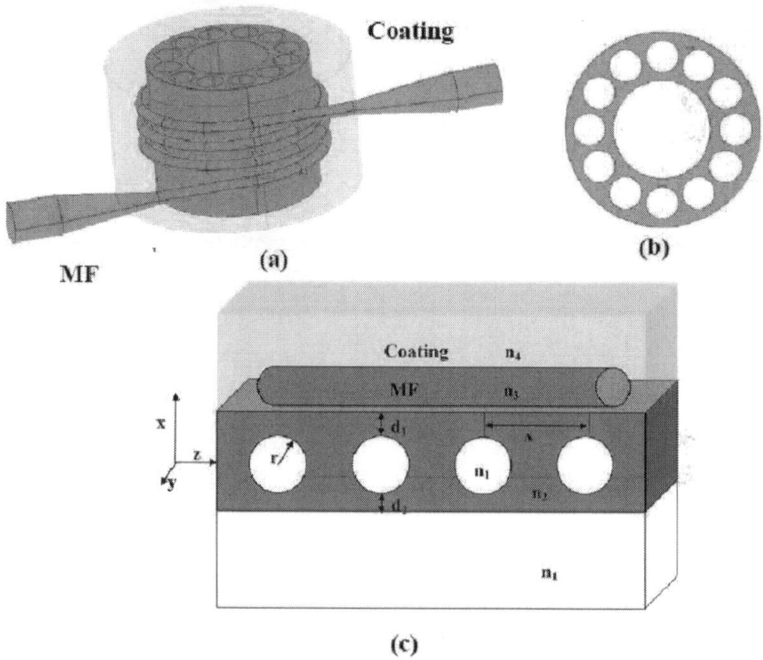

Figure 7. a) Schematic of the embedded SCMG. (b) Cross-section of the support rod (c) the equivalent planar structure, n_1, n_2, n_3, n_4 are the refractive indices of the hole, the rod, the MF, and the coating, respectively[34, 37]. Copyright 2009 Optical Society of America and Copyright 2010 IEEE.

This method is also extremely flexible and can be used to manufacture chirped MF gratings: chirping can be achieved by tuning the hole radius and pitch, or by changing the distance between MF and holes. Since$\lambda=2nneff(deff)\Lambda/m$

, there are three simple ways to realize chirp: by tuning Λ, d = d_1+d_2, and r. Chirp rates are defined as [34, 37]:

$$\begin{cases} \frac{\partial\lambda}{\partial\Lambda} = \frac{2}{m}\frac{\partial n_{neff}}{\partial d_{eff}}\frac{\pi r^2}{\Lambda} + \frac{2}{m}n_{neff} \\ \frac{\partial\lambda}{\partial r} = \frac{2}{m}\frac{\partial n_{neff}}{\partial d_{eff}}(2\Lambda - 2\pi r) \\ \frac{\partial\lambda}{\partial d} = \frac{2}{m}\frac{\partial n_{neff}}{\partial d_{eff}}\Lambda \end{cases} \tag{11}$$

All the derivatives $\partial\lambda/\partial\Lambda$, $\partial\lambda/\partial d$ and $\partial\lambda/\partial r$ depend on $\partial n_{neff}/\partial d_{eff}$, which in turn is strongly dependent on d_{eff} and r. $\partial n_{neff}/\partial d_{eff}$ is of the order of 10^{-2} μm^{-1} and can achieve the maximum at $d_{eff} \sim 1$ μm. For m = 2, from Eq. (11) $\partial\lambda/\partial\Lambda >> \partial\lambda/\partial d \sim \partial\lambda/\partial r$; since $\partial\lambda/\partial\Lambda > 1$, chirped grating can be easily realized by tuning the grating period: current PCF fabrication techniques make the low-cost period-tuning possible. Moreover, $\partial\lambda/\partial d \sim \partial\lambda/\partial r \sim 10^{-5}$: which is enough to produce small precise chirps in a simpler and cheaper way than inducing a temperature, strain or refractive-index gradients [34, 37].

These methods to induce chirp mainly depend on the geometry of the support rod; in fact hole pattern has a periodicity related to the MF turn around the support rod, the circumference of which limits the chirp length. An alternative method to achieve chirped gratings relies on tilting the MF with respect to the support rod longitudinal axis: if the MF cross the rod at an angle φ the Bragg condition is expressed by [42]$\lambda=2neff\Lambda/\cos\varphi/m$; by changing φ gradually, the Bragg wavelength changes gradually, it is possible to control the grating Bragg wavelength and get a chirped grating. When φ is very small, the chirp rate is given by [34, 37]

$$\frac{\partial\lambda}{\partial\varphi} = -2\frac{n_{eff}\Lambda\sin\varphi}{m\cos^2\varphi} \approx 2\frac{n_{eff}\Lambda\varphi}{m} = \lambda\varphi \tag{12}$$

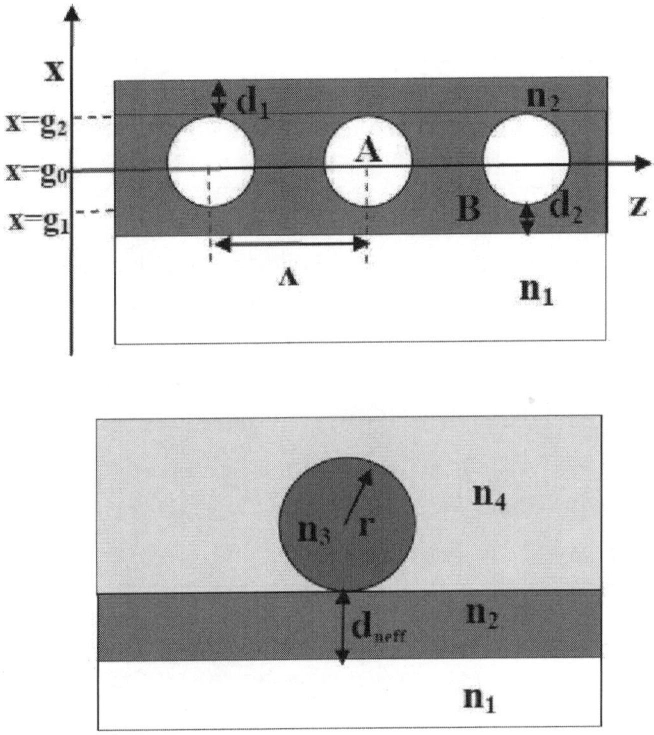

Figure 8. Top: Illustration of the outer layer structure of the support rod in Streifer's theory. n_1 and n_2 represent the refractive indices of the hole and rod, Λ the distance between two adjacent holes, g_1 and g_2 the boundaries between different layers and g_0 represents the new equivalent boundary between n_2 and n_1. Bottom: cross-section of the MF in the equivalent outer straight layer structure; n_3 and n_4 are the indices of the MF and coating; r is the radius of the MF and d_{eff} is the effective wall thickness[34, 37]. Copyright 2010 Optical Society of America and Copyright 2009 IEEE.

The chirp range varies considerably: from 0.3 nm/° at $\varphi = 0.1°$ to 300 nm/° at $\varphi = 10°$. These two chirping methods allow for a great deal of flexibility in the chirp design [34, 37].

The use of microstructured support rods to make gratings provides an extreme flexibility because it is easier to be dealt with than the microfiber itself. By controlling the air holes geometry, it is possible to get several-layer corrugations or phase shifted gratings. If the rod is coated with an active layer or the holes are filled with an active medium, it generates a laser. If holes are used as microfluidic channels, the support rod can work as a sensor. Moreover, since only a very short piece of rod is needed for each device, the average device cost is very low. The device can be coated with stable polymer such as Teflon and UV375 and keep it very strong [34, 37].

A similar method relies on laying the MF on a substrate with pre-treated microstructures [see Fig. 9]. The fraction of power propagating in the evanescent field interacts with the periodically distributed patterns in the rod or the substrate, and light transmission can then be modulated. Both methods avoid post-processing the thin MFs and have great flexibility. However, the MFs have to be coated with low index polymer [43] which means that they are not suitable for high temperature sensing.

Figure 9. Proposed SCMG by laying the MF on a substrate with pre-treated microstructures[44].

Finally, Ding *et al.*[35] combined metal lift-off technology with lithography to produce metallic surface gratings, which provided a high and constant sensitivity to the ambient medium RI, while Phan Huy *et al.* [36] demonstrated an improvement in the sensitivity of RI by making use of the suspended core of a microstructured fiber.

4. SENSING APPLICATIONS

SCMGs can have a number of possible sensing applications as conventional FBGs. Up to date, SCMGs have been used to measure refractive index, temperature and strain/force. With the extreme small size and flexible geometry, SCMGs offer great prospects for developing novel sensors with a very small perturbation on the object being measured.

4.1. Refractive index sensing

Much of the SCMG applications relate to RI sensing because of the very large MF evanescent field. For a typical SCMG sensor immersed in ambient liquid with RI in the range 1.32 – 1.46, S_a varies from 10 nm/RIU (refractive index unit) to 10^3 nm/RIU, according to the MF radius and the ambient liquid sensed, whatever for a FIB-milled or rod-wrapped SCMG. Usually, a smaller radius and a larger ambient medium RI result in a higher sensitivity regardless of the fabrication method. For example, Liang *et al.* got a sensitivity of 16 nm/RIU at a RI around 1.35 with a MF 6 μm in diameter [27] while 660 nm/RIU was reached by Liu *et al.* at a RI of 1.39 by using a 1.8 μm-diameter MF [14]. Both of them agree well with what is predicted from Eq. (5).

In addition to all these nonmetallic SCMGs, metallic gratings have also been proposed for RI sensing. The existence of metal causes light to be coupled to modes of different properties [10, 35]. Figure 10(a) shows the reflection spectra of the metal-dielectric-hybrid SCMG [Fig. 5(d)] immersed in air, acetone, and isopropanol, respectively. The extinction ratio is about ~ 10 dB. The small degree of chirp has been ascribed to the non-uniformity of the taper. The taper diameter difference at the grating extremities is less than 1 μm as illustrated in Fig. 5(d). According to simulations, different effective refractive indices resulting from the variation of the diameter induce resonant wavelength shifts of ~ 4 nm. Some small ripples in Fig. 10(a) can probably be due to a degree of chirping [10].

The grating spectrum presents several valleys and peaks with different characteristics in a 100 nm wide spectral range [10]. The peaks shift when the outer environment changes from acetone to isopropanol. However, these valleys and peaks show larger shifts at longer wavelengths, while those at shorter wavelengths shift much less and almost stop at specific wavelengths, meaning that the reflected light can be coupled to different modes. In the micrometer-diameter metal-dielectric-hybrid fiber tip, several modes are probably excited with close propagation constant because of the metal cladding [45]. Some modes are well confined in the tip and have negligible field overlap with the liquid while other modes are not. The different valleys and peaks correspond to the coupling between these different forward and backward propagating modes, with different response properties to any outer environment change [10].

The metal-dielectric-hybrid grating showed RI sensitive (a in Fig. 10(a)) and insensitive (d in Fig. 10(a)) behavior for different resonant modes [10]. S_a of the sensitive channel (125 nm/RIU) is one order of magnitude larger than that of a nonmetallic SCMG with the same radius whereas S_a of the insensitive channel (8 nm/RIU) is one order of magnitude smaller. This can be attributed to the fact that the introduction of metal film causes the MF to support both surface guided modes (which have a larger modal overlap with the ambient medium, a and b in Fig. 10) and bound modes (where most of the energy is located in the dielectric core, c and d in Fig. 10). The smallest sensitivity can be further decreased to nearly zero by optimizing the tip grating profile and metal coating. Because of many different properties on the outer liquid refractive index, the metal-dielectric-hybrid grating can be applied as a multi-parameter sensor and the index-insensitive channel can be used to simultaneously measure temperature, pressure, and so on [10].

4.2. Temperature sensing

Although thermal post-processing and hydrogen loading have been shown to induce grating capable of standing temperatures as high as 1300 °C in conventional fibers [46], in MF thermometers, up to now, only SCMGs without polymer have been reported operating above 200 °C[9, 13]. The sensitivity of these components is around 20pm/°C, similar to the value predictable using Eq. (6). Figure 11 is the experimental characterization of the FIB milled SCMG demonstrated using the sample shown in Fig. 5(a). As the temperature increases, the Bragg wavelength red shifts. The extremely short SCMG length (~ 36.6μm) and wide operating range (~ 20 – 450°C) presents it as a promising candidate for detecting temperature change in ultra-small space.

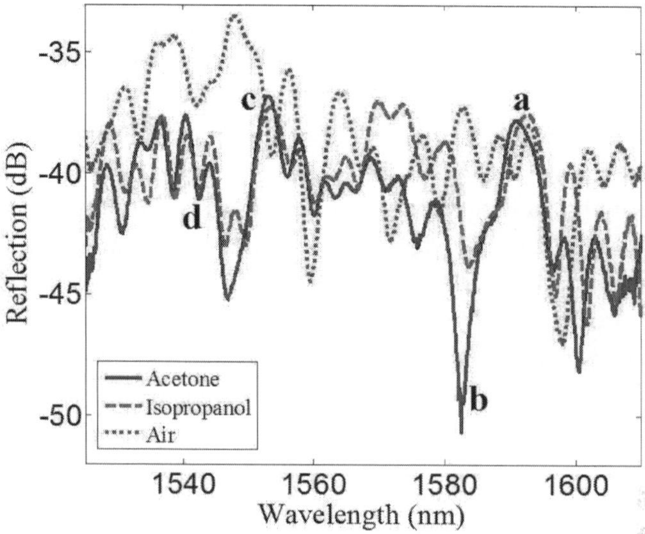

Figure 10. Top: measured reflection spectra of the metal-dielectric-hybrid grating when immersed in air, acetone, and isopropanol. a, b, c, d denote different peaks and valleys labelled. Centre: Dependence of wavelength shift on ambient RI for different modes in a metal-dielectric-hybrid SCMG. Bottom: Calculated effective index of one cladding mode and one core mode as a function of the outer liquid refractive index n_a. The radius of the fiber tip is assumed to be 3 μm with a golden coating 20 nm thick [*10*]. Copyright 2011 IEEE.

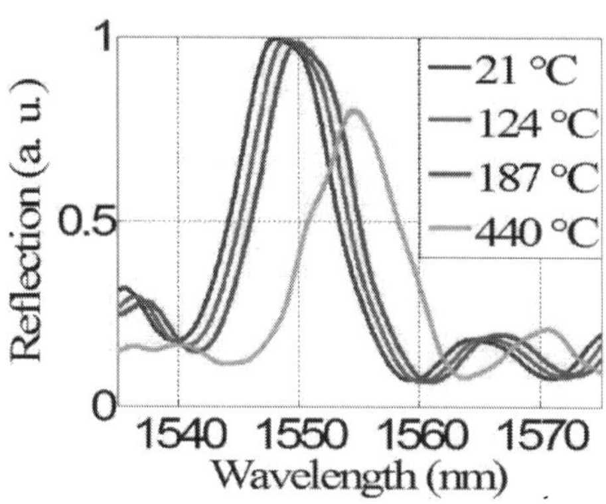

Figure 11. Reflection spectra of the FIB-milled SCMG in air at four different temperatures.

4.3. Strain/Force sensing

Although S_S remains almost the same for different MF diameters[26], S_F varies with the MF radius according to Eq. (9). A SCMG with diameter of 3.5 μm reaches a force sensitivity of ~ 1900 nm/N, which is more than three orders of magnitude compared to that of a conventional fibers [26]. If the Bragg wavelength can be detected to an accuracy of 0.05 nm, forces in the order of 10^{-5} N can be measured. For the sample reported in Fig. 5(d), where the silica constitute only a small fraction of the SCMG cross section, a further three orders of magnitude improvement in sensitivity is predicted, with S_F reaching values in excess of 10^6 nm/N, corresponding to forces of the order of nN. The SCMG strain/force sensors could offer attractive properties monitoring strain/force changes in power plant pipelines, airplane wings, and other civil engineering structures [19].

5. CONCLUSIONS

This review presented the fabrication, operating principles and applications of surface corrugated microfiber Bragg gratings (SCMGs). SCMGs can potentially outperform conventional FBGs because of their large evanescent field and compactness. Taking advantage of their extreme small size and unique geometry, SCMGs exhibit beneficial sensor properties such as fast response, high resolution and small-objection detections. SCMGs could find promising sensing applications in detecting parameter variations in ultra-small space [19].

ACKNOWLEDGEMENT

F. Xu and Y.-q Lu acknowledge the support from National 973 program under contract No. 2011CBA00200 and 2012CB921803, NSFC program No. 11074117 and 60977039, and the Priority Academic Program Development of Jiangsu Higher Education Institutions (PAPD). G. Brambilla gratefully acknowledges the Royal Society (London, U.K.) for his University Research Fellowship.

REFERENCES

1. K. O. Hill, B. Malo, F. Bilodeau, D. C. Johnson, and J. Albert, Bragg gratings fabricated in monomode photosensitive optical fiber by UV exposure through a phase maskAppl. Phys. Lett. 62, 1035 EOF1993

2. S. J. Mihailov, C. W. Smelser, P. Lu, R. B. Walker, D. Grobnic, H. M. Ding, G. Henderson, and J. Unruh, Fiber Bragg gratings made with a phase mask and 800nm femtosecond radiation,Opt. Lett. 28, 995-997 (2003

3. S. A. Slattery, D. N. Nikogosyan, and G. Brambilla, Fiber Bragg grating inscription by high-intensity femtosecond UV laser light: comparison with other existing methods of fabricationJ. Opt. Soc. Am. B-Opt. Phys. 22, 354 EOF361 EOF2005

4. C. Y. Lin, and L. A. Wang, A. Wavelength- And, loss-tunable band-rejection filter based on corrugated long-period fiber grating," IEEE Photon. Technol. Lett. 13, 332 EOF334 EOF2001

5. G. Brambilla, F. Xu, P. Horak, Y. Jung, F. Koizumi, N. P. Sessions, E. Koukharenko, X. Feng, G. S. Murugan, J. S. Wilkinson, and D. J. Richardson, Optical fiber nanowires and microwires: fabrication and applicationsAdv. Opt. Photon. 1, 107-161 (2009

6. T. L. Lowder, K. H. Smith, B. L. Ipson, A. R. Hawkins, R. H. Selfridge, and S. M. Schultz, High-Temperature Sensing Using Surface Relief Fiber Bragg GratingsIEEE Photonic. Technol. Lett. 17, 1926 EOF1928 EOF2005

7. D. Grobnic, S. J. Mihailov, D. Huimin, and C. W. Smelser, Bragg grating evanescent field sensor made in biconical tapered fiber with femtosecond IR radiationIEEE Photon. Technol. Lett. 18, 160 EOF2006

8. X. Fang, C. R. Liao, and D. N. Wang, Femtosecond laser fabricated fiber Bragg grating in microfiber for refractive index sensingOpt. Lett. 35, 1007-1009 (2010

9. J. -l. Kou, S. -j. Qiu, F. Xu, and Y. -q. Lu, Demonstration of a compact temperature sensor based on first-order Bragg grating in a tapered fiber probe,Opt. Express 19, 18452 EOF7 EOF2011

10. J. -l. Kou, S. -j. Qiu, F. Xu, Y. -q. Lu, Y. Yuan, and G. Zhao, Miniaturized metal-dielectric-hybrid fiber tip grating for refractive index sensingIEEE Photon. Technol. Lett. 23, 1712 EOF1714 EOF2011

11. M. Ding, P. Wang, T. Lee, and G. Brambilla, A microfiber cavity with minimal-volume confinementAppl. Phys. Lett. 99, 051105 (2011

12. M. Ding, M. N. Zervas, and G. Brambilla, A compact broadband microfiber Bragg grating,Opt. Express 19, 15621 EOF6 EOF2011

13. J. Feng, M. Ding, J. -l. Kou, F. Xu, and Y. -q. Lu, An optical fiber tip micrograting thermometerIEEE Photon. J. 3, 810 EOF814 EOF2011

14. Y. Liu, C. Meng, A. P. Zhang, Y. Xiao, H. Yu, and L. Tong, Compact microfiber Bragg gratings with high-index contrast,Opt. Lett. 36, 3115 EOF7 EOF2011

15. K. P. Nayak, F. Le, Y. Kien, K. Kawai, K. Hakuta, H. T. Nakajima, Miyazaki, and Y. Sugimoto, "Cavity formation on an optical nanofiber using focused ion beam milling technique,Opt. Express 19, 14040 EOF50 EOF2011

16. F. L. Kien, K. P. Nayak, and K. Hakuta, Nanofibers with Bragg gratings from equidistant holesJ. Mod. Opt. 59, 274 EOF286 EOF2012

17. K. Handa, S. Peng, and T. Tamir, Improved perturbation analysis of dielectric gratingsApplied PhysicsA: Materials Science & Processing 5, 325 EOF328 EOF1975

18. T. Erdogan, Fiber grating spectraIEEE J. Lightwave Technol. 15, 1277 EOF1294 EOF1997

19. J. -L. Kou, M. Ding, J. Feng, Y. -Q. Lu, F. Xu, and G. Brambilla, Microfiber-Based Bragg Gratings for Sensing Applications: A Review,Sensors 12, 8861 EOF76 EOF2012

20. F. Xu, P. Horak, and G. Brambilla, Optimized Design of Microcoil ResonatorsJournal of Lightwave Technology1561 EOF1567 EOF2007

21. F. Xu, P. Horak, and G. Brambilla, Optical microfiber coil resonator refractometric sensor: erratum,Opt. Express 15, 9385 EOF2007

22. I. M. White, H. Zhu, J. D. Suter, N. M. Hanumegowda, H. Oveys, M. Zourob, and X. Fan, Refractometric sensors for lab-on-a-chip based on optical. ring resonators," Ieee Sensors Journal28 EOF35 EOF2007

23. M. Adams, G. A. Derose, M. Loncar, and A. Scherer, Lithographically fabricated optical cavities for refractive index sensingJournal of Vacuum Science & Technology B 23, 3168-3173 (2005

24. J. -l. Kou, J. Feng, L. Ye, F. Xu, and Y. -q. Lu, Miniaturized fiber taper reflective interferometer for high temperature measurement,Opt. Express 18, 14245 EOF50 EOF2011

25. L. Sang-mae, S. S. Saini, and J. Myung-yung, Simultaneous measurement of refractive Index, temperature, and strain using etched-core fiber Bragg gratingsensors," IEEE Photon. Technol. Lett. 22, 1431-1433 (2010

26. T. Wieduwilt, S. Bruckner, and H. Bartelt, High force measurement sensitivity with fiber Bragg gratings fabricated in uniform-waist fiber tapersMeas. Sci. Technol. 22, 075201 EOF2011

27. W. Liang, Y. Huang, Y. Xu, R. K. Lee, and A. Yariv, Highly sensitive fiber Bragg grating refractive index sensorsAppl. Phys. Lett. 86, 151122 (2005

28. A. Iadicicco, S. Campopiano, A. Cutolo, M. Giordano, and A. Cusano, Refractive index sensor based on microstructured fiber Bragg gratingIEEE Photon. Technol. Lett. 17, 1250 EOF1252 EOF2005

29. A. Iadicicco, A. Cusano, A. Cutolo, R. Bernini, and M. Giordano, Thinned fiber Bragg gratings as high sensitivity refractive index sensorIEEE Photon. Technol. Lett. 16, 1149 EOF1151 EOF2004

30. R. Yang, J. Long, T. Yan-nan, S. Li-peng, L. Jie, and G. Bai-ou, High-efficiency ultraviolet inscription of Bragg gratings in microfibersIEEE Photon. J. 4, 181 EOF186 EOF2012

31. Y. Ran, Y. -N. Tan, L. -P. Sun, S. Gao, J. Li, L. Jin, and B. -O. Guan, 193nm excimer laser inscribed Bragg gratings in microfibers for refractive index sensingOpt. Express 19, 18577 EOF2011

32. F. Bilodeau, B. Malo, J. Albert, D. C. Johnson, K. O. Hill, Y. Hibino, M. Abe, and M. Kawachi, Photosensitization of optical fiber and silica-on-silicon/silica waveguides,Opt. Lett. 18, 953 EOF5 EOF1993

33. Y. Zhang, B. Lin, S. C. Tjin, H. Zhang, G. Wang, P. Shum, and X. Zhang, Refractive index sensing based on higher-order mode reflection of a microfiber Bragg grating,Opt. Express 18, 26345 EOF50 EOF2010

34. F. Xu, G. Brambilla, and Y. Lu, A microfluidic refractometric sensor based on gratings in optical fibre microwires,Opt. Express 17, 20866 EOF71 EOF2009

35. W. Ding, S. R. Andrews, T. A. Birks, and S. A. Maier, Modal coupling in fiber tapers decorated with metallic surface gratingsOpt. Lett. 31, 2556-2558 (2006

36. M. C. Phan, G. Huy, V. Laffont, P. Dewynter, P. Ferdinand, J. Roy, L. Auguste, D. Pagnoux, W. Blanc, and B. Dussardier, Three-hole microstructured optical fiber for efficient fiber Bragg grating refractometer,Opt. Lett. 32, 2390 EOF2 EOF2007

37. F. Xu, G. Brambilla, J. Feng, and Y. -Q. Lu, A microfiber Bragg grating based on a microstructured rod: a proposalIEEE Photon. Technol. Lett. 22, 218 EOF220 EOF2010

38. R. R. Gattass, and E. Mazur, Femtosecond laser micromachining in transparent materials," Nat. Photon. 2, 219-225 (2008

39. J. C. Knight, Photonic crystal fibres,Natur 424, 847 EOF51 EOF2003

40. M. Bayindir, F. Sorin, A. F. Abouraddy, J. Viens, S. D. Hart, J. D. Joannopoulos, and Y. Fink, Metal-insulator-semiconductor optoelectronic fibres,Natur 431, 826 EOF9 EOF2004

41. G. Brambilla, F. Xu, and X. Feng, Fabrication of optical fibre nanowires and their optical and mechanical characterisationElectron. Lett. 42, 517 EOF518 EOF2006

42. W. Streifer, and A. Hardy, Analysis of two-dimensional waveguides with misaligned or curved gratingsQuantum Electronics, IEEE Journal of 14, 935 EOF943 EOF1978

43. F. Xu, and G. Brambilla, Preservation of Micro-Optical Fibers by EmbeddingJpn. J. Apl. Phys. 47, 6675 EOF2008

44. J. -l. Kou, Z. -d. Huang, G. Zhu, F. Xu, and Y. -q. Lu, Wave guiding properties and sensitivity of D-shaped optical fiber microwire devicesApplied Physics B 102, 615 EOF619 EOF2011

45. J. -L. Kou, S. -J. Qiu, F. Xu, Y. -Q. Lu, Y. Yuan, and G. Zhao, Miniaturized Metal-Dielectric-Hybrid Fiber Tip Grating for Refractive Index SensingIEEE Photon. Technol. Lett. 23, 1712 EOF1714 EOF2011

46. J. Canning, M. Stevenson, S. Bandyopadhyay, and K. Cook, Extreme Silica Optical Fibre GratingsSensors 8, 6448-6452 (2008

CHAPTER 2

High Temperature Optical Fiber Sensor Based on Compact Fattened Long-Period Fiber Gratings

Ruth I. Mata-Chávez [1,*], Alejandro Martínez-Rios [2], Julián M. Estudillo-Ayala [3], Everardo Vargas-Rodríguez [1], Roberto Rojas-Laguna [3], Juan C. Hernández-García [3], Ana D. Guzmán-Chávez [1], David Claudio-González [1] and Eduardo Huerta-Mascotte [1]

[1] Espintrónica y Óptica Aplicada-CA, Departamento de Estudios Multidisciplinarios, DICIS, Universidad de Guanajuato, Av. Universidad s/n, Col. Yacatitas, Yuriria, Gto, 38940, Mexico

[2] Centro de Investigaciones en Óptica, Loma del Bosque 115, Col. Lomas del Campestre, León, Guanajuato, 37150, Mexico

[3] Optoelectrónica-CA, Departamento de Electrónica, DICIS, Universidad de Guanajuato, Carr. Salamanca-Valle Km 3.5 + 1.8, Palo Blanco, Salamanca, Gto, 36855, Mexico

ABSTRACT

A compact high temperature fiber sensor where the sensor head consists of a short fattened long period fiber grating (F-LPFG) of at least 2 mm in length and background loss of −5 dBm is reported. On purpose two different F-LPFGs were used to measure temperature variations, taking advantage of their broad spectrum and the slope characteristics of the erbium light source. This approach affected the spectrum gain as the linear band shifting took place. The measured sensitivity of the long period fiber gratings were about 72 pm/°C in a range from 25 to 500 °C. Here, the temperature rate of the experiment was 0.17 °C/s and the temperature response time was within 3 s. Moreover, temperature changes were detected with an InGaAs photodetector, where a sensitivity of 0.05 mV/°C was achieved.

KEYWORDS

long-period fiber grating; temperature sensor; electric arc; dispersion shifted fiber; fattening

1. INTRODUCTION

Optical fiber devices compete with conventional mechanic and electronic sensors due to their physical advantages such as compact size, light weight, immunity to electromagnetic interference, high sensitivity and temperature resistance [1–3]. There is a simple and low cost fabrication procedure to fabricate fiber gratings, the arc discharge method, which is applied to any type of fiber using a fusion splicer machine. The standard fabrication method varies the quantity of tension applied over the fiber and leads to tapered gratings [4,5]. Theory and functionality of these optical devices is completely known, as well as their sensitivities, which depend on fiber type, the grating period and the external perturbation [6,7]. These types of LPFG are well suited for many sensing applications but sometimes their length of more than 10 cm is a great disadvantage. Long period gratings in two and three layered optical fibers, as well as photonic crystal LPFG had been proposed for temperature measurements using electric discharges [8–13]. Studies of thermal behavior have been reported for LPFG with different core dopants, which were fabricated with electric arc applying the tension method and a period of 540 μm for up to 1,200 °C [14]. Rego et al. presented a study in LPFGs thermal behavior on different types of optical fibers, including Corning DSF with this method [15]. Sakata et al. proposed an intensity-based fiber sensing technique where temperature sensing was made by intensity detection by employing a combination of a light-emitting diode (LED) and a photodetector [16].In 2008, the fattening method was proposed, as an alternative method, to fabricate F-LPFG.A commercial fusion splicer machine and a dispersion shifted fiber were used and approximately3 mm in length gratings were achieved [17]. A F-LPFG is a LPFG formed by enlarging the fiber diameter applying several arc discharges in the same region. Periods of minimum of 100 nm can be applied since there is a geometrical contribution induced through the fattening or compression of the fused sections which changes the dimensions of the fiber structure, i.e., the core and cladding dimensions will increase. This geometrical change will modify the effective indices of the interacting modes as in the case of tapered fiber where the geometrical change affects the wave-guiding properties.

We present an enhanced temperature characterization in short F-LPFGs for sensing implementation comparing with results of our previous work. This is, a spectral shift of ~50 pm/°C at a temperature range from 25 °C to 190 °C was obtained with fattened positions separated by ~25 nm, where at each position the fusion process is repeated five times resulting in a total overlap distance of ~55 μm. A low power white light source was used [17]. In this work two individual F-LPFGs with 1,524 and 1,546 nm central dip wavelengths were fabricated. Temperature variations were characterized in transmission using an EDF light source as the excitation source. Two different methods of detection for the output were used, an OSA and InGaAs photodetector. The measuring range of temperature was of 25 to500 °C and the sensitivities calculated by each method were of 72 pm/°C and 0.05 mV/°C, respectively. Using advantageously the slope characteristics of the fluorescence spectrum of erbium, optional temperature measurements can be realized with these LPFGs.

2. PRINCIPLE OF THE FATTENED-LONG PERIOD FIBER GRATING

An important parameter of a long-period fiber grating is known as the resonance wavelength and occurs when the phase matching condition between the core and the m-th cladding mode is satisfied. This condition can be written as:

$$\lambda_{res} = \left[n_{coe}(\lambda_{\upsilon}) - n^{\upsilon}_{cle}(\lambda_{\upsilon})\right]\Lambda \qquad (1)$$

where Λ is the grating period, υ is the mode number and λ_{res} is the resonance wavelength of υ-th order. $n_{coe}(\lambda_\upsilon)$ and $\eta vcle(\lambda v)$ are the effective indices of the core mode and the cladding mode, respectively. The modal effective index can be calculated by solving the characteristic equation obtained by considering a four-layered waveguide as the dispersion shifted fiber [17]. As temperature LPFG sensors, the temperature sensitivity of the resonance wavelength can be expressed as [11]:

$$\frac{d\lambda_\upsilon}{dT} \cong (C_{co} - C_{cl})(\eta_{01} - \eta_{0m})\Lambda\gamma \qquad (2)$$

with:

$$\gamma = \frac{1}{1 - \Lambda\left(\frac{dN_{01}}{d\lambda} - \frac{dN_{0m}}{d\lambda}\right)}$$

where η_{01} and η_{0m} denote the fractional powers in the core for the LP_{01} and LP_{0m} modes. γ describes the effect of the modal dispersion evaluated at the resonance wavelength, and C_{co}, C_{cl}, N_{01} and N_{0m} are the thermo-optic coefficients and the effective indexes of core and cladding, respectively [9].

The small size of the fattened grating contributes to the size of the bandwidth of the stopband measured at full width at half maximum [6], this is:

$$\Delta\lambda = \frac{0.8\lambda^2}{L(n_{eff} - n_{cl})} \qquad (3)$$

where λ is the resonance wavelength, n_{eff} is the effective index of the guided LP_{01} mode, L is the grating length and n_{cl} is the cladding refractive index. Figure 1 shows the bandwidth variation in aF-LPFG at 1,546 nm with its length. Shorter length gratings show broad spectra with higher slopes in the tails of the resonant band. Launching conditions do not change with temperature, so it is possible to measure these variations as intensity modulation with a wavelength laser source slightly off the resonance central wavelength.

The graphs were numerically obtained from Equation (3). The assumed parameters were estimated for a DSF using the BPM RSoft® (Ossining, NY, USA) which results were based on the radii values:a = 2.9 μm, b = 5 μm and c = 8 μm, and the refractive index values n_1 = 1.4598, n_2 = 1.4498, n_3 = 1.4498 and n_4 = 1 of which the effective indices were obtained for different wavelengths.Of course these values might change for different types of fiber structures and refractive indexes so it is not an accurate indicator for LPFG lengths but illustrates the approximated size of the stopband measured at full width at half maximum.

We have to take into account that the electric arc is a variable which depends on external surrounding factors as humidity, temperature and the metal electrodes deterioration. Using an average effective index as for example Δn_{eff} = 0.015 and fixing the optical wavelength, different graphics can be obtained to serve as a guide for bandwidths obtained for an approximated F-LPFG length.

3. FATTENED LONG-PERIOD GRATING FABRICATION

Using the electric discharge method with a fusion splicer machine (Fitel S175) in manual mode, the diameter of at least 2 mm length dispersion-shifted fiber (DSF) was enlarged by applying several electric arcs at the same point. The experimental setup for monitoring the evolution of the F-LPFG was a white light source coupled to one end of the DSF and the other end to the OSA. Splice discharge parameters that are mentioned in reference [16,17], were used to fabricate these particular gratings. The process was repeated five times along ~2 mm of fiber length, until filtering functions were observed in the OSA [18]. Figure 2(a) shows a picture of the amplification of a fattened LPFG where the spaced glass lobes are observed and Figure 2(b) shows a scale in mm with the same F-LPFG of about 2 mm in length. Figure 2(c) shows that neither the lobes maintain the original diameter size of 125 μm nor the smallest diameter between the lobes.

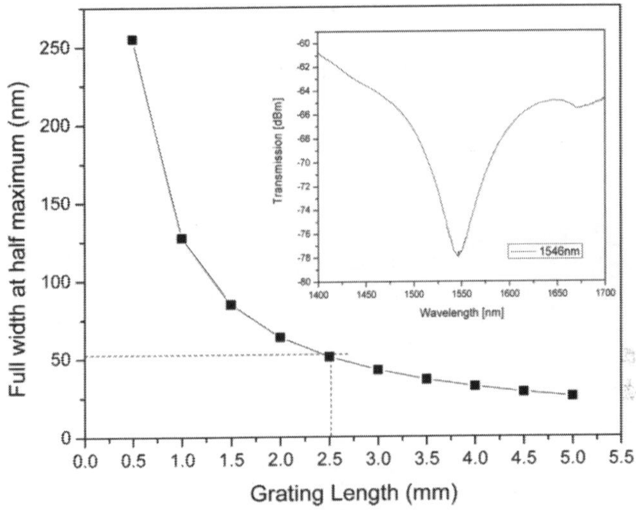

(a)

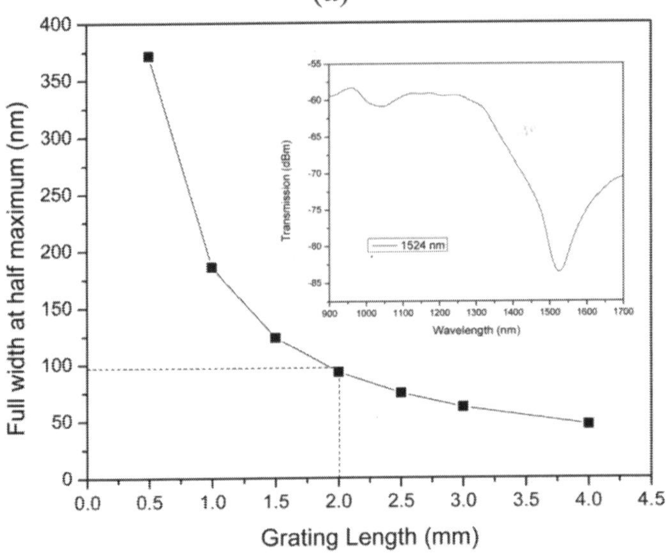

(b)

Figure 1. (**a**) Influence of the size of the F-LPFG with the dip bandwidth. Inset shows a 1,546 nm F-LPFG spectra al FWHM of about 51 nm. (**b**) A F-LPFG around 2 mm in length at 1,524 nm with a FWHM spectra of 90.43 nm.

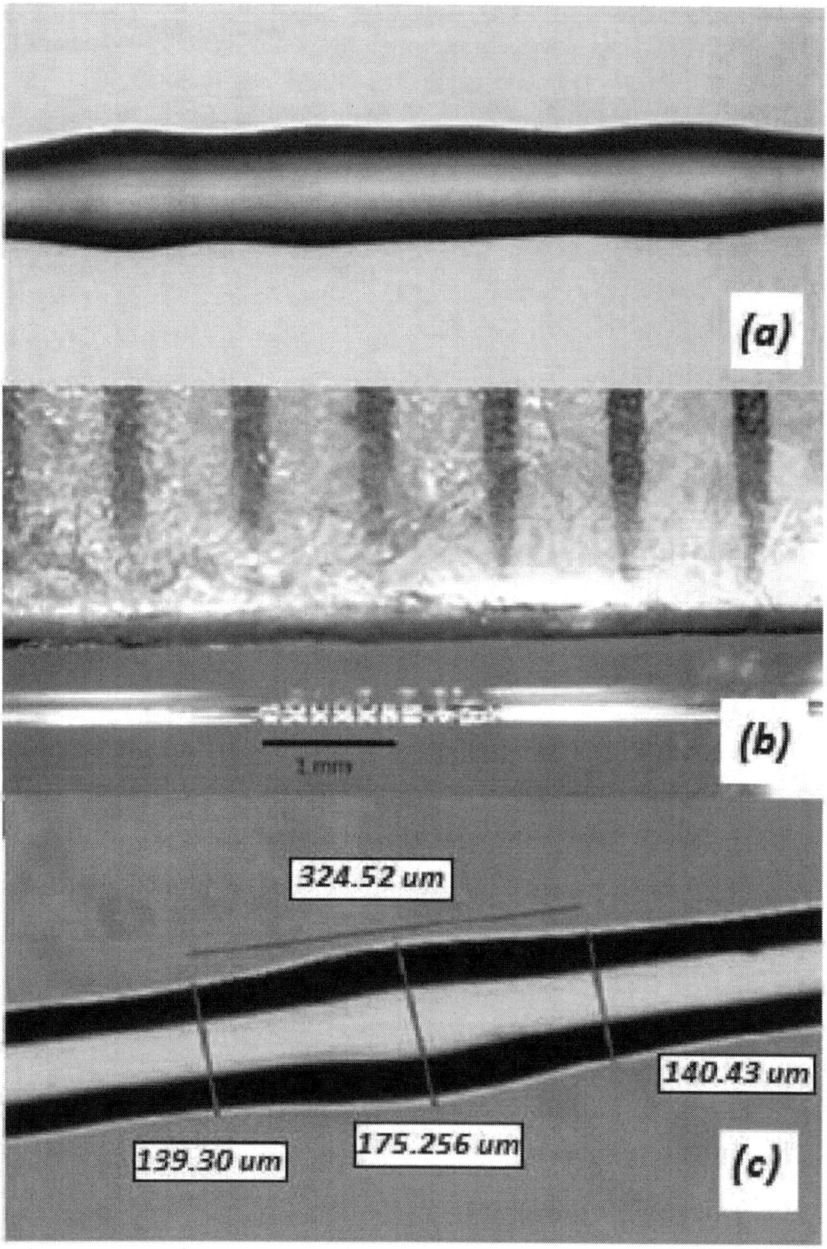

Figure 2. (**a**) Photograph of a F-LPFG in DS Fiber (**b**) F-LPFG with ~2 mm length (**c**) DSF with fattened diameter with a maximum enlargement of ~175 μm.

The horizontal line represents the length of the fattened grating. This change in the whole fiber structure makes possible the coupling of modes to observe filtering functions at transmission spectra. Preliminary work has been done in order to find the fiber cladding modes for the F-LPFG. Using an ideal structure of the dispersion shifted fiber and the beam propagation method (BPM) of RSoft® Software, the structure radii values were included as: $a = 2.9$ μm, $b = 5$ μm and $c = 8$ μm, as well as the refractive index values $n_1 = 1.4598$, $n_2 = 1.4498$, $n_3 = 1.4498$ and $n_4 = 1$ [18]. With this method we have found that the modes that are excited in the fiber grating, depending on the wavelength, are basically LP01 and LP11, as well as LP21 and LP03. We consider that these results are still not accurate due to the four layered structure. This is, the DSF has a core and a ring core with a depressed section between them and the rest is cladding. The analysis assumes an ideal single mode structure whose results cannot be said to be exact, but can be used as a guideline as a satisfactory method is found to obtain improved results.

This F-LPFG was fabricated with a grating period of 150 μm to achieve a central dip wavelength ~1,546 nm. We must take into account the influence of the finite size of the electric arc, about 300 μm, over the affected surface area of the optical fiber. This makes it sometimes difficult to fabricate controlled central dip wavelength LPFGs with smaller grating period values. Although the strong geometric variation formed over the optical fiber structure makes possible to achieve a LPFG with a minimum of four grating periods. The F-LPFGs were obtained with a loss dip ~15 dB, resonance wavelengths of 1,546 nm ± 20 nm and a broad bandwidth of more than 100 nm at full top width as observed in the inset of Figure 1.

Varying the number of arc discharges and alternating the grating period from 100 μm to 150 μm, a LPFG with a resonance wavelength at 1,524 nm was obtained. There is no accurate method to select the central wavelength in the fabrication process at this point, but we have observed that depends on the manufacturing process. Actually we work in the process optimization to reduce the background losses by fixing and enlarging the grating period and to find a form to select the exact location of the resonance band without losing its compactness. However, what we do is to change the program values as arc power, arc duration, and pre-fusion times as well as the number of arc discharges. That is, if we want to fabricate a F-LPFG around 1,550 nm, we apply five arc discharges with arc power of 85 mW, 650 ms for arc duration and 250 ms of pre-fusion time. After applying the five arc discharges the fiber is displaced 150 μm and start applying other five discharges. The obtained grating is about 3 mm in length with bandwidths around 55 nm. With this approach it has been possible to manufacture F-LPFG with central wavelengths around 1,300 nm near the cutoff wavelength, but it has been more difficult to manufacture them below the cutoff wavelength.

Opposed to the standard LPFG fabrication by electric discharge [15], this alternative method does not use a dead weight to apply axial tension over the optical fiber and it is not necessary to unscrew a fixture to place the fiber. An advantage over the UV method is that the grating is not erased or optically degraded with exposures to high temperatures (<200 °C). The hot push delay and the overlapping distance contribute to the grating formation and these values are advantageously established when manipulating the splicer machine in manual mode. Background losses can be reduced by applying arc discharges in manual mode which are of –5 dBm, and this is an advantage of the fabrication method. The fabrication process modifies the refractive index of the glass, giving the grating the filtering characteristics depicted in the inset of Figure 1(a,b). With the fattening method by electric arc, compact devices are always achieved and the advantage is that bulky and expensive equipment can be avoided for sensing interrogation, achieving in general, a whole compact sensing device. However, the fabrication of LPFG with electric arc is still not very accurate. Resonance wavelengths might result around ±5 nm for periods below 200 μm.

4. FATTENED LONG-PERIOD GRATING CHARACTERIZATION AND RESPONSE

The experimental setup shown in Figure 3 was used to study the temperature effect over the transmission spectrum of the F-LPFG. The characterization system consists on a fluorescence light source, which is obtained by pumping an EDF with a laser diode (LD) at wavelength 980 nm through a wavelength division multiplexer (WDM 980/1,550 nm). An isolator was spliced to the common port of the WDM and the EDF which was employed to prevent any reflected light. Then the output of the source, another end of the EDF, was spliced to the F-LPFG (head sensor). The F-LPFG was put over a temperature controller with a range from 25 to 550 °C. Finally the output sensor could be coupled either to the OSA (spectral resolution 20 pm) or the InGaAs photodetector (PD)-oscilloscope.

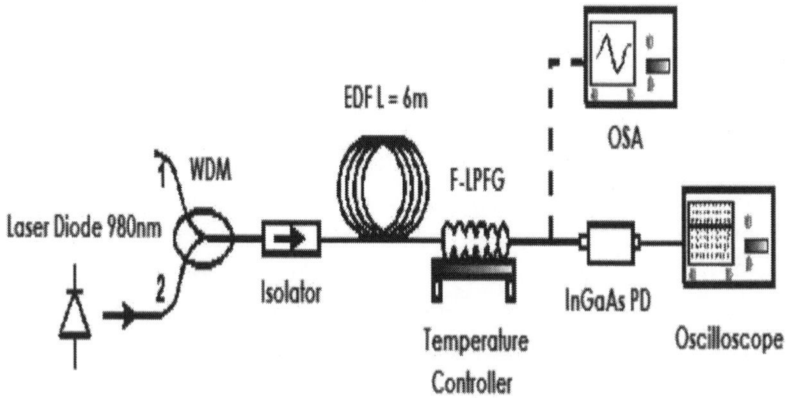

Figure 3. Sketch of the F-LPFG temperature sensor characterization system.

The spectra recorded with the OSA; in log scale, shows a common linear wavelength shift response as temperature increases for F-LPFGs at 1,524 and 1,546 nm central dip wavelengths, Figure 4.

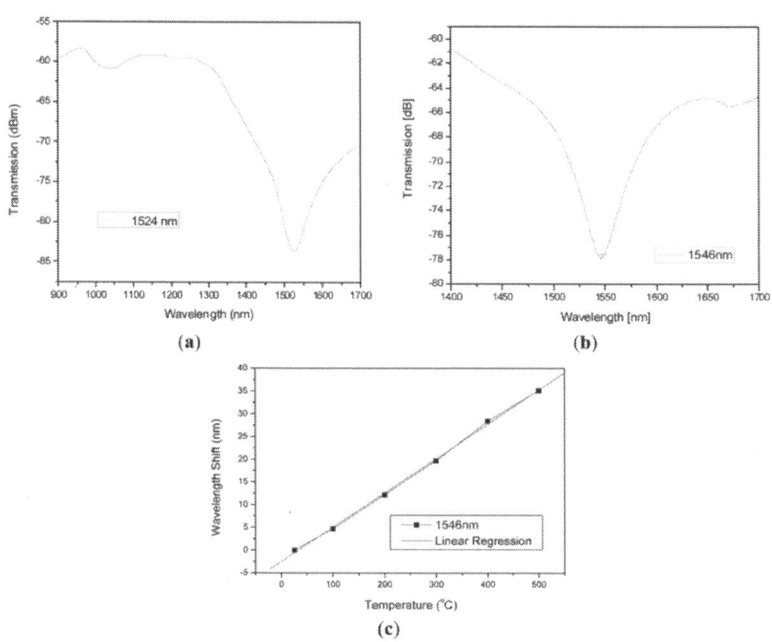

Figure 4. (**a**) Transmission spectra of 1,524 nm F-LPFG, (**b**) Transmission spectra of 1,546 nm F-LPFG, (**c**) Linear regression for 1,546 nm F-LPFG.

The sensitivity achieved was ~72 pm/°C, for the case when the experimental temperature rate was 0.17 °C/s, which is a greater value than the ones obtained in previous works with DS F-LPFG at1,550 nm (where the fiber was fattened at positions separated by ~25 μm) [17]. The transmission spectrum is similar to almost all LPFGs. The attenuation band shifts to longer wavelengths while the depth of the band is attenuated as the temperature rises. The evolution of the isolation loss with temperature in a range from (25–500 °C) is about −1.5 dBm. These effects have been observed and reported in many literature or experiments with LPFGs. The repeatability was measured several times. The temperature was increased to the limit of 500 °C and then to low temperature. The fiber was exposed 10 min at increments of temperature of 100 °C. There were small variations during the characterizations which are illustrated in Figure 5. The overall temperature response time measured was within 3 s for this sensor.

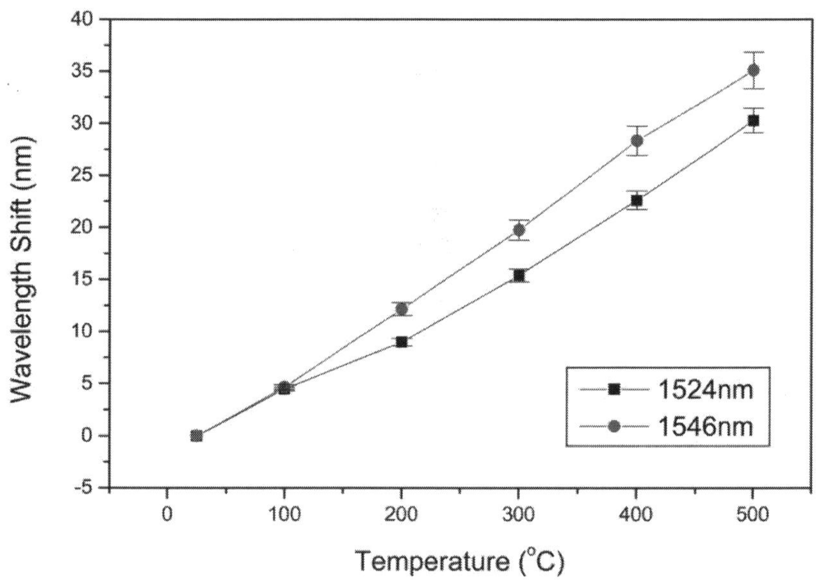

Figure 5. Transmission spectra of both F-LPFG with variations of maximum of 5%.

If we wanted to build an electronic device to read the movement of the band, so it would be more complicated to detect the central wavelength. For this reason the change of logarithmic to the linear regime to which, instead of reading the displacement, power changes can be read in intensity as the resonance band changes with temperature.

Measurements of light power variations of F-LPFGs with temperature could be registered with the OSA in the linear scale. The fluorescence spectrum of erbium in the wavelength range underneath ~1,532 nm has a positive slope and wavelengths over 1,540 nm has a negative slope [19], for this reason the increase of the wavelength of the F-LPFG that is centered at 1,546 nm when increasing temperature caused an increase on the power measurements and the ones centered at 1,524 nm caused a decrease on the power measurements. Transmission spectra for F-LPFG at 1,524 nm and 1,546 nm with an erbium source are shown in Figure 6.

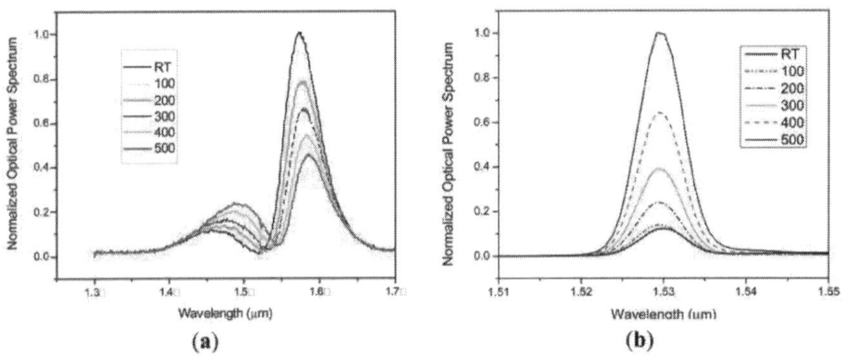

Figure 6. (**a**) Power spectrum of a 1,524 nm F-LPFG, (**b**) Power spectrum of a 1,546 nm F-LPFG.

As part of the power spectrum is filtered with the 1,524 nm F-LPFG, it keeps enough power which decreases as the temperature increases (Figure 6(a)). This is the effect that occurs when placing a LPFG near the left end of the fluorescence spectrum. The F-LPFG at central wavelength of 1,546 nm immediately attenuates the power spectrum and when temperature increases, from room temperature to 500 °C, produces the increment of power as depicted in the Figure 6(b). These measurements were made with a maximum source power of 50 mW. After observing the power behavior of the F-LPFG with the erbium source, the goal is to measure the F-LPFG output as a variable easy to read by any electronic device as an oscilloscope or electronic homemade board that can help to characterizein volts/°C and achieve a very compact temperature sensor. Figure 7 shows these steady changes as temperature varies every ±100 °C. The highest sensitivity registered in volts is 0.05 mV/°C.The curves in Figure 7 are not linear. This effect can be avoided using wider and flattened spectral sources. Some candidates are photoluminescent diodes with a high emission intensity, Erbium doped fibers with wide flattened spectrum can also be used or a supercontinuum wide spectrum white light source.

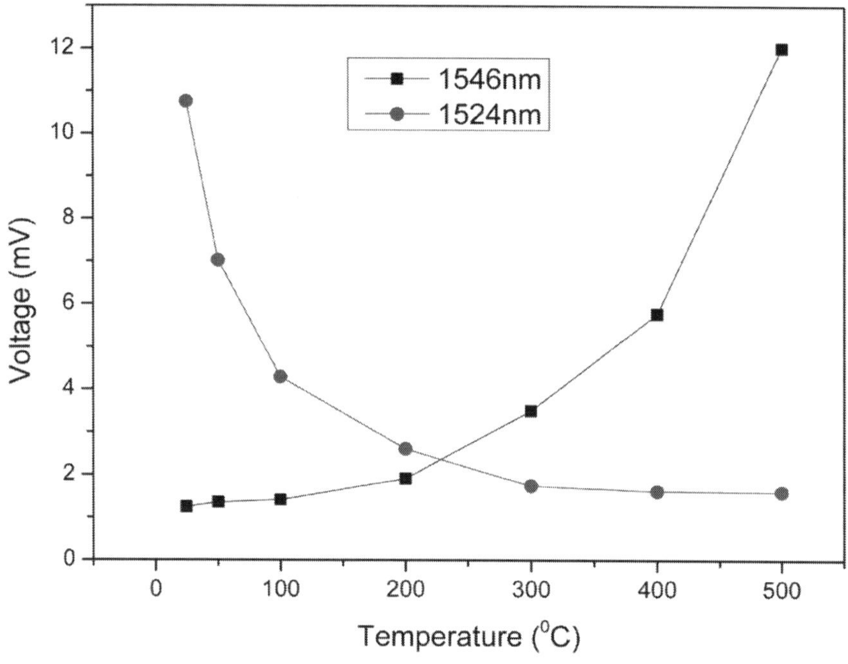

Figure 7. Raise and fall of voltage due to temperature changes of 1,524 nm and 1,550 nm F-LPFGs.

One of the main advantages of the current sensor is that in contrast with other long period fiber gratings fabricated by electric arc, CO_2 or others, this device is totally insensitive to changes in the external refractive index and has very low sensitivity to bending compared with other long period gratings where the length may be in excess of 1 cm [6,20]. In the case of LPGs fabricated by UV, the application of high temperature may erase the gratings in most cases. Hence for this particular application the F-LPFG has a clear advantage with respect to UV LPFG of similar lengths, and with respect to other arc-induced LPFG has the advantage of lower sensitivity to external factors like refractive index and bending, although it has temperature sensitivity in the same order. Furthermore, the OSA can be replaced by a small power meter or any compact electronic device capable to convert light into voltage or data which can enormously reduce the size of a complete optical sensor.The optical device results very compact due to the compact grating and the compact electronic device.

5. CONCLUSIONS

Compact fattened DS-LPFGs can be fabricated with lengths of ~2 mm with a dynamic wavelength range ~36 nm for a temperature range from 25 to 550 °C. The measured sensitivity of 72 pm/°C improves the sensing devices fabricated with F-LPFG. The advantage of getting a LPFG with a very broad spectrum is that, with steep spectra sources, sensors whose behavior will depend on the position of the central dip wavelengths, can be deployed. In this way, it is easier to obtain the signal sensitivity (i.e., voltage) which in this case turned out to be 0.05 mV/°C. The small sensor based on these shortF-LPFGs has potential applications where smaller devices are required without loss of performance and reliability as in ovens, medicine, bio-sensing, optical communications, and actual opto-mechatronics systems in industry.

ACKNOWLEDGMENTS

Author Mata-Chávez thanks CONACyT for the retention grant No. 93956 and project # 166361.To CONCyTEG for its support with project No. 09-16-K662-084-P04. We also thank DAIP-UG and PROMEP-SEP for its partial support with different projects on F-LPFG.

REFERENCES

1. Bhatia, V. Applications of long period gratings and multi-parameter sensing. Opt. Express. **1999**, 14, 457–466.

2. Culshaw, B. Optical fiber sensor technologies. Oportunities and perhaps-pitfalls. J. Lightwave Technol. **2004**, 22, 39–50.

3. Cusano, A.; Iadiccico, A.; Pilla, P.; Contessa, L.; Campopiano, S.; Cutolo, A.; Giordano, M.; Guerra, G.J. Coated long-period fiber greatings as high-sensitivity optochemical sensors. J. Lightwave Technol. **2006**, 24, 1776–1786.

4. Hwang, I.K.; Yun, S.H.; Kim, B.Y. Long period fiber gratings based on periodic microbends. Opt. Lett. **1999**, 24, 1263–1265.

5. Morishita, K.; Miyake, Y. Fabrication and resonance wavelengths of long-period gratings written in a pure-silica photonic crystal fiber by the glass structure change. J. Lightwave Technol. **2004**, 22, 625–630.

6. Vengsarkar, A.M.; Lemaire, P.J.; Judkins, J.B.; Bhatia, V.; Erdogan, T.; Sipe, J.E. Long period fiber gratings as band rejection filters. J. Lightwave Technol. **1996**, 14, 58–65.

7. James, S.W.; Tatam, R. Optical fibre long-period grating sensors: Characteristics and application. Meas. Sci. Technol. **2003**, 14, R49–R61.

8. Humbet, G.; Malki, A. Characterizations at very high temperature of electric arc-inducedlong-period fiber gratings. Opt. Commun. **2002**, 208, 329–335.

9. Ng, M.N.; Chiang, K.S. Thermal effects on the transmission spectra of long-period fiber gratings. Opt. Commun. **2002**, 208, 321–327.

10. Allsop, T.; Webb, D.J.; Bennion, I. Investigations of the spectral sensitivity of long period gratings fabricated in three-layered optical fiber. J. Lightwave Technol. **2003**, 21, 264–268.

11. Chaubey, S.; Joshi, P.; Kumar, M.; Arya, R.; Nath, A.K.; Kher, S. Design and development of long period gratings sensors for temperature monitoring. Sãdhanã **2007**, 32, 513–519.

12. Petrovic, J.S.; Dobb, H.; Mezentsev, V.K.; Kalli, K.; Webb, D.; Bennion, I. Sensitivity of LPFs in PCF fabricated by an electric arc to temperature, strain and external refractive index. J. Lightwave Technol. **2007**, 25, 1306–1312.

13. Zhao, C.L.; Demokan, M.S.; Jin, W.; Xiao, L. A cheap and practical FBG temperature sensor utilizing a long period grating in a photonic crystal fiber. Opt. Commun. **2007**, 276, 242–245.

14. Rego, G.; Okhotnikov, O.; Dianov, E.; Sulimov, V. High-temperature stability of long period fiber gratings produced using an electric arc. J. Lightwave Technol. **2001**, 19, 1574–1579.

15. Rego, G.; Marques, P.V.S.; Santos, J.L.; Salgado, H.M. Arc-induced long-period gratings. Fiber Int. Opt. **2005**, 24, 245–259.

16. Hajime, S.; Hisashi, I. Optical fiber temperature sensor using a pair of nonidentical long period fiber gratings for intensity based sensing. Opt. Commun. **2007**, 280, 87–90.

17. Mata-Chávez, R.I.; Martinez-Rios, A.; Torres-Gomez, I.; Alvarez-Chavez, J.A.; Selvas-Aguilar, R.; Estudillo-Ayala, J.M. Wavelength band-rejection filtres based on opticalfiber fattening by fusion splicing. Opt. Laser Technol. **2008**, 40, 671–675.

18. Mata-Chávez, R.I.; Martinez-Rios, A.; Torres-Gomez, I.; Selvas-Aguilar, R.; Estudillo-Ayala, J.M. Mach-Zehnder all-fiber interferometer using two in-series fattened fiber gratings. Opt. Rev. **2008**, 15, 230–235.

19. Guzman-Chávez, A.D.; Barmenkov, O.; Kir, Y. Spectral dependence of the excited-state absorption of erbium in silica fiber within the 1.48–1.59 μm range. Appl. Phys. Lett. **2008**, 92, 191111–191113.

20. Chan, H.M.; Alhassen, F.; Tomov, I.V.; Lee, H.P. Fabrication and mode identification of compact long-period gratings written by CO_2 laser. IEEE. Photo. Technol. Lett. **2008**, 20, 611–613.

CHAPTER 3

A Guide to Fiber Bragg Grating Sensors

Marcelo M. Werneck[1], Regina C. S. B. Allil[1, 2], Bessie A. Ribeiro[1] and Fábio V. B. de Nazaré[1]

[1] Instrumentation and Photonics Laboratory, Electrical Engineering Program, Universidade Federal do Rio de Janeiro (UFRJ), RJ, Brazil

[2] Division of Chemical, Biological and Nuclear Defense, Biological Defense Laboratory, Brazilian Army Technological Center (CTEx) RJ, Brazil

1. INTRODUCTION

Optical fiber sensors (OFS) appeared just after the invention of the practical optical fiber by Corning Glass Works in 1970, now Corning Incorporated, that produced the first fiber with losses below 20 dB/km. At the beginning of this era, optical devices such as laser, photodetectors and the optical fibers were very expensive, afforded only by telecom companies to circumvent the old saturated copper telephone network. With the great diffusion of the optical fiber technology during the 1980's and on, optoelectronic devices became less expensive, what favored their use in OFS.

OFS can be applied in many branches of the industry but we will concentrate here the electrical power industry. In this area, the operators need to measure and monitor some important physical parameters that include:

- Strain (με)
- Vibration of structures and machines
- Electric current (from A to kA)
- Voltage (from mV to MV)
- Impedance (μΩ)
- Leakage current of insulators (μA to mA)
- Temperature
- Pressure
- Gas concentration
- Distance between stationary and rotating or moving parts

In the electrical power industry (EPI) we have two facts that can cause collapse of an electronic sensor: presence of high voltage and presence of high electromagnetic interference. Therefore, depending on where we want to measure a parameter it can be very difficult or even impossible to use a conventional sensor. The best option to circumvent this is through the use of an OFS, because the fiber is made of dielectric material sand, therefore, it is possible to place them very close or even over a high potential conductor and they do not necessarily need electrical power at the sensor location.

Another problem with conventional sensors is that they all need electric energy to work. However, providing electric energy at the sensor location is sometimes difficult if the device needs to be far away from any appropriated power supply. It happens in long high voltage transmission lines, at high voltage potentials, along pipe-lines or in deep ocean, for instance. Since OFS are passive sensors they do not need electric energy to work.

Therefore we can mention some, very specific characteristics of OFS that are well exploited when applied to the EPI:

- High immunity to Electromagnetic Interference (EMI)

- Electrical insulation
- Absence of metallic parts
- Local electrical power not required
- Lightweight and compactness
- Easy maintenance
- Chemically inert even against corrosion
- Work over long distances
- Several sensors can be multiplexed on the same fiber

There are many options to develop an OFS. The easiest way is by making the measurement to modulate the light amplitude that is the power, and ending up with an amplitude modulated sensor. These sensors were very common at the beginning of OFS era but they gradually were substituted by wavelength based sensors. These are more stable and self-calibrated as the wavelength does not depend on losses due connectors, modal drifts, macro bends, or LED and LASER ageing/drifts.

In this Chapter we will concentrate on a very special type of OFS: the Fiber Bragg Grating (FBG) sensors.

2. THEORY AND MODELS OF FBG

Fiber Bragg Grating (FBG) technology is one of the most popular choices for optical fiber sensors for strain or temperature measurements due to their simple manufacture, as we will see later on, and due to the relatively strong reflected signal. They are formed by a periodic modulations of the index of refraction of the fiber core along the longitudinal direction and can be produced by various techniques. The term fiber Bragg grating was borrowed from the Bragg law and applied to the periodical structures inscribed inside the core of conventional telecom fiber. Therefore, before entering the theory of fiber Bragg grating itself, it is worth to go back one century behind in order to review the Bragg law.

Sir William Lawrence Bragg, was born in 1890, a British physicist and X-ray crystallographer, was the discoverer, in 1912, of the Bragg law of X-ray diffraction. This principle is used until today for the study and determination of crystal structure, particularly in thin film research. Sir Bragg, together with his father, won the Nobel Prize for Physics in 1915 for an important step in the development of X-ray crystallography.

Bragg diffraction occurs for an electromagnetic radiation whose wavelength is the same order of magnitude of the atomic spacing, when incident upon a crystalline material. In this case the radiation is scattered in a specular fashion by the atoms of the material and experiences constructive interference in accordance to Bragg's law. For a crystalline solid with lattice planes separated by a distance *d* the waves are scattered and interfere constructively if the path length of each wave is equal to an integer multiple of the wavelength. Figure 1 shows the idea. Bragg's law describes the condition for constructive interference from several crystallographic planes of the crystalline lattice separated by a distance *d*:

$$2d\sin\theta = n\lambda \tag{1}$$

Where θ is the incident angle, n is an integer and λ is the wavelength. A diffraction pattern is obtained by measuring the intensity of the scattered radiation as a function of the angle θ. Whenever the scattered waves satisfy the Bragg condition it is observed a strong intensity in the diffraction pattern, known as Bragg peak.

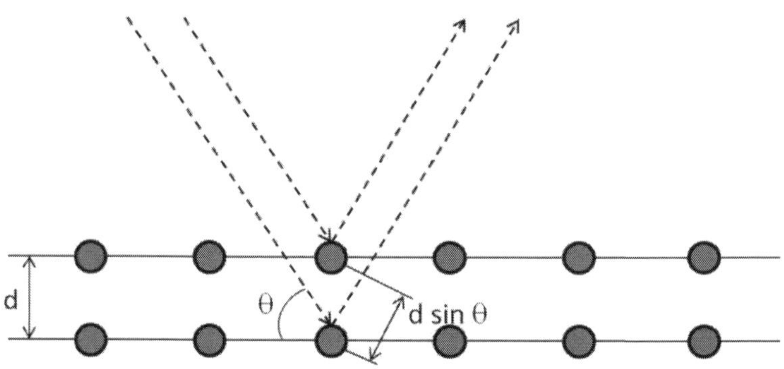

Figure 1. An incident radiation is reflected by the lattice structure of a crystal and will interfere constructively if the Bragg law is obeyed.

The first observations of index of refraction changes were noticed in germane silicate fibers and were reported by Kenneth Hill and co-workers in 1978 [1]. They described a permanent grating written in the core of the fiber by an argon ion laser line at 488 nm launched into the fiber by a microscope objective. This particular grating had a very weak index modulation, resulting in a narrow-band reflection filter at the writing wavelength. In reality, this phenomenon happened by chance, when they injected a high power blue light into de fiber, unexpectedly, after a few minutes, the transmitted light decayed. It was Friday but they were so puzzled with this phenomenon that Hill returned to his laboratory on Saturday to make a new experiment. He wanted to know where the light was going to and he had a clue. He used a thin microscope slide as a beam splitter in order to monitor a possible reflection from the fiber and there was the missing light [2]. The explanation is that at the end of the fiber about 4% of the light was reflected by Fresnel reflection which, in its way backwards, interfered with the ongoing light producing an interference pattern. This pattern contained peaks and valleys of a stationary wave which imprinted permanently the pattern into the core of the fiber as an index of refraction modulation. Initially, the reflected light intensity is low, but after some time, it grows in intensity until almost all the light launched into the fiber is back-reflected. The growth in back-reflected light was explained in terms of a new effect called “photosensitivity”.

After the inscription of the grating into de fiber’s core, due the periodic modulation of the index of refraction, light guided along the core of the fiber will be weakly reflected by each grating plane by Fresnel effect. The reflected light from each grating plane will join together with the other reflections in the backward direction. This addition may be constructive or destructive, depending whether the wavelength of the incoming light meets the Bragg condition of Eq. (1).

Now, according to Eq. (1), since θ=90° and *d* is the distance between peaks of the interference pattern, λ=2d for n=1 is the approximate wavelength of the reflection peak. That is, the fiber now acts as a dichroic mirror, reflecting part of the incoming spectrum. Equation (1), developed for vacuum, has to be adapted for silica, since the distances traveled by light are affected by the index of refraction of the fiber:

$$\lambda_B = 2n_{eff}\Lambda \qquad (2)$$

Therefore the Bragg wavelength (λ_B) of an FBG is a function of the effective refractive index of the fiber (η_{eff}) and the periodicity of the grating (Λ).

The photosensitivity phenomenon in optical fibers remained unexplored for several years after its discovery, mainly due the fact that the resulted Bragg wavelength was always a function of the wavelength of the light source used and very far away from the interested region of the spectrum, namely, the third telecommunication window. However, a renewed interest appeared years later with the demonstration of the side writing technique by Gerry Meltz and Bill Morey of United Technology Research Center [3] and later on with the possibility of tuning the Bragg wavelength into the C Band of the telecom spectrum.

Equation (2), also known as the Bragg reflection wavelength, is the peak wavelength of the narrowband spectral component reflected by the FBG. The FWHM (full-width-half-maximum) or bandwidth of this reflection depends on several parameters, particularly the grating length. Typically, the FWHM is 0.05 to 0.3 nm in most sensor applications. Figure 2 shows a typical Bragg reflection peak. The lateral lobes sometimes pose problems in automatic identification of the center wavelength and in telecom applications, such as wavelength division multiplexing (WDM), these side-lobes need to be suppressed in order to reduced the separation between the optical carriers, according to ITU-T-G.694.1 (International Telecommunications Union). The side-lobes can be suppressed during the FBG fabrication by a technique known as apodization.

From Eq. (2) we see that the Bragg wavelength only depends on the distance between gratings (Λ) and the effective index of refraction (n_{eff}).

Essentially, any external agent that is capable of changing Λ will displace the reflected spectrum centered at Bragg wavelength. A longitudinal deformation, due to an external force, for instance, may change both Λ and n_{eff}, the latter by the photo-elastic effect and the former by increasing the pitch of the grating. Equally, a variation in temperature can also change both parameters, via thermal dilation and thermo-optic effect respectively.

Therefore FBG is essentially a sensor of temperature and strain but, by designing the proper interface, many other measurements can be made to impose perturbation on the grating resulting in a shift in the Bragg wavelength which can then be used as a parameter transducer. Therefore, by using an FBG as a sensor we can obtain measurements of strain, temperature, pressure, vibration, displacement, etc.

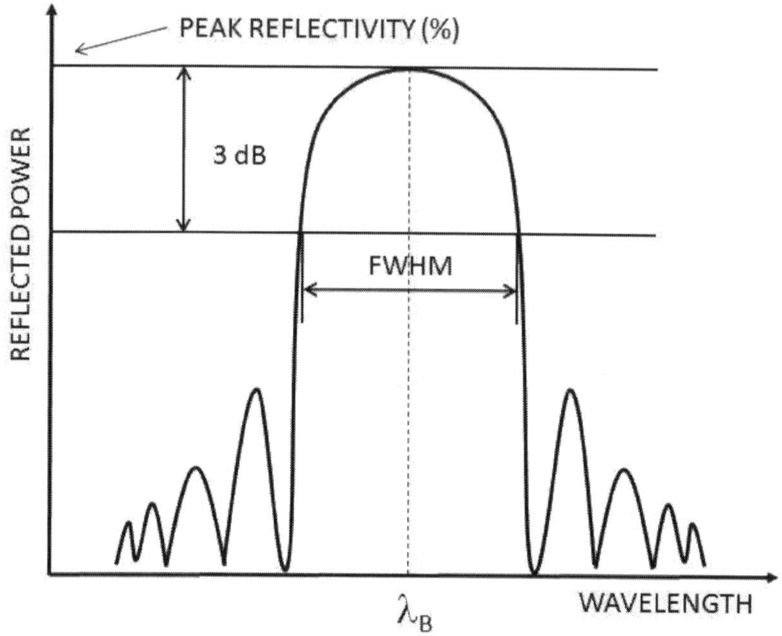

Figure 2. A typical Bragg reflection wave shape with its parameters defined.

Besides the influence of temperature and strain on the Bragg grating periodicity, one can also use n_{eff}, the fiber effective refractive index (RI) as a parameter transducer. The effective refractive index is an average of the RI of the core (n_{co}) and the RI of the cladding (n_{clad}) of the fiber. This parameter depends on how much the evanescent field of the core penetrates into the cladding. Since the fiber cladding diameter (125 μm) is much larger than the evanescent field, the effective RI is undisturbed by external influences. However by a corrosion of the fiber cladding by acid etching, one can reach the evanescent field which lies about 1.5 μm from the core interface. Now, the effective RI depends also on the surrounding RI, that is, the air, a gas or a liquid outside the fiber and we just created a device that can measure the RI of substances.

Since the strain or temperature measurements are encoded into wavelength shifts, these sensors are also self-calibrated because wavelength is an absolute parameter. Thus these sensors do not drift on the total light levels, losses in the connecting fibers and couplers or light source power. Additionally, the wavelength encoded nature of the output also allows the use of wavelength division multiplexing technique (WDM) by assigning each sensor to a different wavelength range of the available light source spectrum.

Using such a device and by injecting a spectrally broadband source of light into the fiber, a narrowband spectral component at the Bragg wavelength will be reflected by the grating. This spectral component will be missed in the transmitted signal, but the remainder of this light may be used to illuminate other FBGs in the same fiber, each one tuned to a different Bragg wavelength. The final result of such an arrangement is that we will have all Bragg peak reflections of each FBG back at the beginning of the fiber, each one in its specific wavelength range.

In order to calculate the sensitivity of the Bragg wavelength with temperature and strain we start from Eq. (2) and notice that the sensitivity with temperature is the partial derivative with respect of temperature:

$$\frac{\Delta\lambda_B}{\Delta T} = 2n_{eff}\frac{\partial\Lambda}{\partial T} + 2\Lambda\frac{\partial n_{eff}}{\partial T} \tag{3}$$

Substituting twice (2) in (3) we get:

$$\frac{\Delta\lambda_B}{\Delta T} = \frac{1}{\Lambda}\frac{\partial\Lambda}{\partial T}\lambda_B + \frac{1}{n_{eff}}\frac{\partial n_{eff}}{\partial T}\lambda_B$$

or rearranging,

$$\frac{\Delta\lambda_B}{\lambda_B} = \frac{1}{\Lambda}\frac{\partial\Lambda}{\partial T}\Delta T + \frac{1}{n_{eff}}\frac{\partial n_{eff}}{\partial T}\Delta T$$

The first term is the thermal expansion of silica (α) and the second term is the thermo-optic coefficient (η) representing the temperature dependence of the refractive index (dn/dT). Substituting we have:

$$\frac{\Delta\lambda_B}{\lambda_B} = (\alpha + \eta)\Delta T \quad (4)$$

The sensitivity with strain is the partial derivative of (2) with respect to displacement:

$$\frac{\Delta\lambda_B}{\Delta L} = 2n_{eff}\frac{\partial\Lambda}{\partial L} + 2\Lambda\frac{\partial n_{eff}}{\partial L} \quad (5)$$

Substituting twice (2) in (5), we have:

$$\frac{\Delta\lambda_B}{\lambda_B} = \frac{1}{\Lambda}\frac{\partial\Lambda}{\partial L}\Delta L + \frac{1}{n_{eff}}\frac{\partial n_{eff}}{\partial L}\Delta L \quad (6)$$

The first term in Eq. (6) is the strain of the grating period due to the extension of the fiber. Suppose we have a length L of a fiber with an inscribed FBG in it. If we apply a stress on the fiber of ΔL then we will have an relative strain ΔL/L. At the same time if the FBG has a length L_{FBG} it will experience a strain $\Delta L_{FBG}/L_{FBG}$ but since the FBG is in the fiber, then $\Delta L_{FBG}/L_{FBG}=\Delta L/L$. Since the Bragg displacement with extension equals the displacement of the grating period with the same extension and, therefore, the first term in Eq. (6) is the unit.

The second term in Eq. (6) is the photo-elastic coefficient (ρ_e), the variation of the index of refraction with strain. In some solids, depending on the Poisson ratio of the material, this effect is negative, that is, when one expands a transparent medium, as an optical fiber for instance, the index of refraction decreases due to the decrease of density of the material. Then, when an extension is applied to the fiber, the two terms in Eq. (6) produce opposite effects, one by increasing the distance between gratings and thus augmenting the Bragg wavelength and the other by decreasing the effective RI and thus decreasing the Bragg wavelength. The combined effect of both phenomena is the classical form of the Bragg wavelength displacement with strain:

$$\frac{\Delta\lambda_B}{\lambda_B} = (1 - \rho_e)\varepsilon_z \tag{7}$$

where ε_z is the longitudinal strain of the grating. Combining (4) and (7) together we finally end up with the sensitivity of the Bragg wavelength with temperature and strain:

$$\frac{\Delta\lambda_B}{\lambda_B} = (1 - \rho_e)\varepsilon_z + (\alpha + \eta)\Delta T \tag{8}$$

The parameters in (8) have the following values for a silica fiber with a germanium doped core:

ρ_e=0.22,

α=0.55 x 10-6/°C,

And

η=8.6 x 10-6/°C.

Thus the sensitivity of the grating to temperature and strain at the wavelength range of 1550 nm, after substituting the constants in (8) are:

$$\frac{\Delta\lambda_B}{\Delta T} = 14.18 \text{ pm/°C} \tag{9}$$

and:

$$\frac{\Delta\lambda_B}{\Delta\varepsilon} = 1.2 \text{ pm}/\mu\varepsilon \tag{10}$$

These theoretical values, though, are not absolute as each FBG of the same fabrication batch will present slightly different sensitivities, as we will see later in the following sections.

3. TEMPERATURE COMPENSATION

Equation (8) shows that the Bragg displacement is a function of both strain and temperature. By observing only $\Delta\lambda_B$ one cannot tell if the displacement was due to strain, temperature or both. If one wants to measure only temperature, the FBG must be protected against strain which can be simply done by loosely inserting the FBG into a small-bore rigid tubing. However, if one wants to measure strain, it is very difficult to stop variation of local temperature to reach the FBG; instead, we have to compensate this variation. In order to do this we have to measure the local temperature, by a thermistor, for instance, and apply Eq. (4) to calculate the effect of temperature alone in the Bragg wavelength displacement. Then, the displacement of the Bragg wavelength due to strain alone is the total displacement observed minus the displacement due to temperature alone.

This approach is only valid if it was possible to electrically measure the temperature, which is not always the case since the local of interest could be a high voltage environment or a place with a high EMI.

The more elegant way is by the use of another FBG on the same fiber, protected against strain and at the same temperature as its neighbor. The two FBGs will be in the same fiber-optic and will provide two different Bragg reflections, one dependent on strain and temperature and the other dependent only on temperature, for compensation.

From Eq. (3) we have for the first FBG:

$$\Delta\lambda_{B1} = K_{\varepsilon 1}\Delta\varepsilon + K_{T1}\Delta T \tag{11}$$

Where

$$K_{\varepsilon 1} = (1 - \rho_e)\lambda_{B1} \tag{12}$$

$$K_{T1} = (\alpha + \eta)\lambda_{B1} \tag{13}$$

Similarly, for the other FBG we have:

$$\Delta\lambda_{B2} = K_{\varepsilon 2}\Delta\varepsilon + K_{T2}\Delta T \tag{14}$$

Where,

$$K_{\varepsilon 2} = (1 - \rho_e)\lambda_{B2} \tag{15}$$

$$K_{T2} = (\alpha + \eta)\lambda_{B2} \tag{16}$$

But since this FBG is strain free, the first term of (14) will not exist and $K_{\varepsilon 2}$ equals zero. Equations (11) and (14) can be written in matrix form:

$$\begin{bmatrix} \Delta\lambda_{B1} \\ \Delta\lambda_{B2} \end{bmatrix} = \begin{bmatrix} K_{\varepsilon 1} & K_{T1} \\ K_{\varepsilon 2} & K_{T2} \end{bmatrix} \times \begin{bmatrix} \Delta\varepsilon \\ \Delta T \end{bmatrix} \tag{17}$$

Equation (17) is called the wavelength shift matrix because its solution gives us the wavelength displacements of both FBGs as a function of temperature and strain. However, we need to find the sensing matrix that gives us the strain and temperature as a function of the wavelength displacement of each FBG. So, we multiply both sides of Eq. (17) by the inverse of the 2x2 matrix and get to:

$$\begin{bmatrix} \Delta\varepsilon \\ \Delta T \end{bmatrix} = \begin{bmatrix} K_{\varepsilon 1} & K_{T1} \\ K_{\varepsilon 2} & K_{T2} \end{bmatrix}^{-1} \times \begin{bmatrix} \Delta\lambda_{B1} \\ \Delta\lambda_{B2} \end{bmatrix} \tag{18}$$

Inverting the 2x2 matrix we have the sensing matrix:

$$\begin{bmatrix} \Delta\varepsilon \\ \Delta T \end{bmatrix} = \frac{1}{K_{\varepsilon 1}K_{T2} - K_{\varepsilon 2}K_{T1}} \begin{bmatrix} K_{T2} & -K_{T1} \\ -K_{\varepsilon 2} & K_{\varepsilon 1} \end{bmatrix} \times \begin{bmatrix} \Delta\lambda_{B1} \\ \Delta\lambda_{B2} \end{bmatrix} \tag{19}$$

In (19) we notice that if

$$K_{\varepsilon 1}K_{T2} \approx K_{\varepsilon 2}K_{T1} \tag{20}$$

then we would not have a possible solution for Eq. (19) because equations (11) and (14) would be two almost parallel lines. This would happen, for instance, if the two FBGs had the same coefficients and Bragg wavelength reflection and would, therefore displace equally. Notice that Eq. (12) and Eq. (15), as well as Eq. (13) and Eq. (16), respectively, differ only by the Bragg wavelength. So, to avoid the redundancy in Eq. (19) we can use FBGs with Bragg reflections wide apart.

Now we can solve Eq. (19) for strain and temperature:

$$\Delta\varepsilon = \frac{1}{K_{\varepsilon 1}K_{T2}-K_{\varepsilon 2}K_{T1}}(K_{T2}\Delta\lambda_{B1} - K_{T1}\Delta\lambda_{B2}) \tag{21}$$

$$\Delta T = \frac{1}{K_{\varepsilon 1}K_{T2}-K_{\varepsilon 2}K_{T1}}(K_{\varepsilon 1}\Delta\lambda_{B2} - K_{\varepsilon 2}\Delta\lambda_{B1}) \tag{22}$$

Equation (21) gives the real strain of FBG 1 as measured by $\Delta\lambda_{B1}$, compensated against temperature variation measured by $\Delta\lambda_{B2}$. Equation (22) gives the temperature of the sensors. It can be used for further compensation, as for instance the thermal dilation of the metallic parts of the setup.

4. CALIBRATION OF FBG WITH TEMPERATURE AND UNCERTAINTY ASSESSMENT

As it will be seen below, Eq. (9) is not an exact model for the FBG behavior under temperature variation and therefore each FBG has to be independently calibrated in order to be possible to tell the temperature by the Bragg wavelength. In this section we demonstrate the procedure to calibrate an FBG chain made of five FBGs. In this study five FBGs were submitted to temperature variations between 20°C and 85°C in order to verify and quantify the parameters of Eq. (4) [4].

In order to measure temperature, we can use as many FBG as necessary, in different Bragg wavelengths; the only precaution is that each FBG's spectrum should not overlap with its neighbor during its displacement when the temperature varies. To obtain the largest range for five FBGs we distributed them along the available range of most FBGs interrogators, that is, 1530 nm-1570 nm.

The setup used to calibrate the FBGs is shown in Figure 3. The dotted square represents the optical system comprised of a commercial Bragg Meter

(Spectral Eye 400 from FOS&S) that consists of an ASE (Amplified Spontaneous Emission) broadband source used to illuminate the FBGs via Port 1 of the optical circulator. The reflection spectrum of the FBGs returns through Port 2 and is directed via Port 3 to an embedded OSA where the reflected spectrum is detected and measured. All controls and data can be accessed by a computer connected to the USB port of the interrogator.

Figure 4 shows superimposed spectra of five FBGs recorded in a temperature variation from 20°C to 85°C.

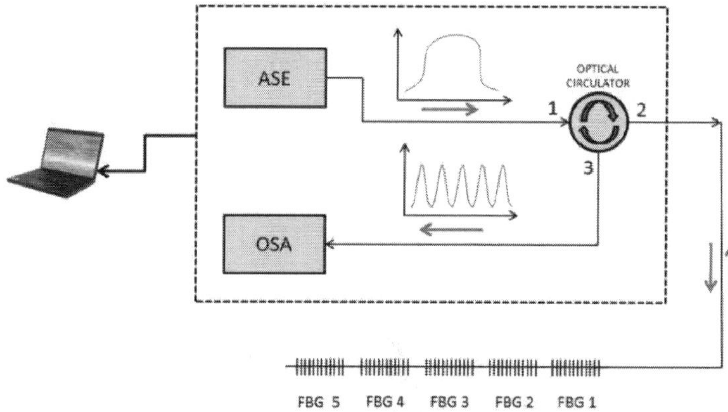

Figure 3. Schematic diagram of the measurement technique [**4**].

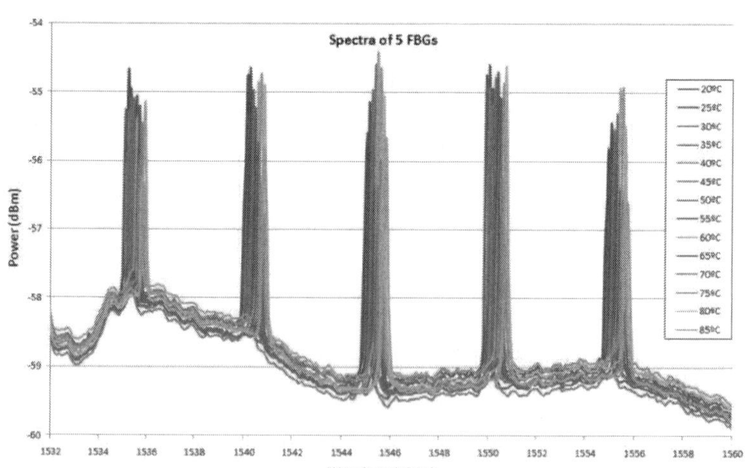

Figure 4. Superimposed spectra of five FBGs recorded in a temperature variation from 20°C to 85°C[**4**].

The procedure to calibrate the sensors followed the sequence: they were immersed simultaneously into a controlled temperature bath and the Bragg wavelengths were monitored and recorded along with the temperature given by a NIST-traceable thermometer (TD 990, Thermolink, 0.1°C resolution and ±1°C accuracy). Five sets of measurements were performed for each sensor in the range of 20°C to 85°C. Table I shows Bragg shift for each temperature and for each FBG.

From the data in Table 1 it is possible to calculate the sensitivity of each sensor, as predicted by (4) and the accuracy of the measurement chain. The graph in Figure 5 was built from the data in Table 1.

Table 2 shows a summary of the calibration parameters: the theoretical and experimental sensitivities, the correlation coefficients of the curve fittings, the root mean square errors (RMSE) and maximum residual errors.

Table 1 Average Bragg center wavelength of each FBG under temperature variation [4]

T (°C)	Average Bragg wavelength peak (nm)				
	FBG1	FBG2	FBG3	FBG4	FBG5
25	1536,001	1540,928	1545,833	1550,819	1555,723
30	1536,067	1540,995	1545,903	1550,888	1555,793
35	1536,128	1541,059	1545,971	1550,959	1555,865
40	1536,183	1541,116	1546,03	1551,015	1555,922
45	1536,245	1541,177	1546,092	1551,083	1555,987
50	1536,310	1541,247	1546,163	1551,147	1556,055
55	1536,368	1541,308	1546,224	1551,213	1556,119
60	1536,427	1541,368	1546,284	1551,273	1556,178
65	1536,497	1541,440	1546,358	1551,345	1556,253
70	1536,557	1541,501	1546,420	1551,408	1556,317
75	1536,618	1541,566	1546,484	1551,469	1556,383
80	1536,680	1541,627	1546,549	1551,539	1556,447

Figure 6 shows the error analysis for FBG1; the maximum positive error was approximately 0.004°C at 30°C. The temperature error measurements for other sensors were within the range of ±0.007°C.

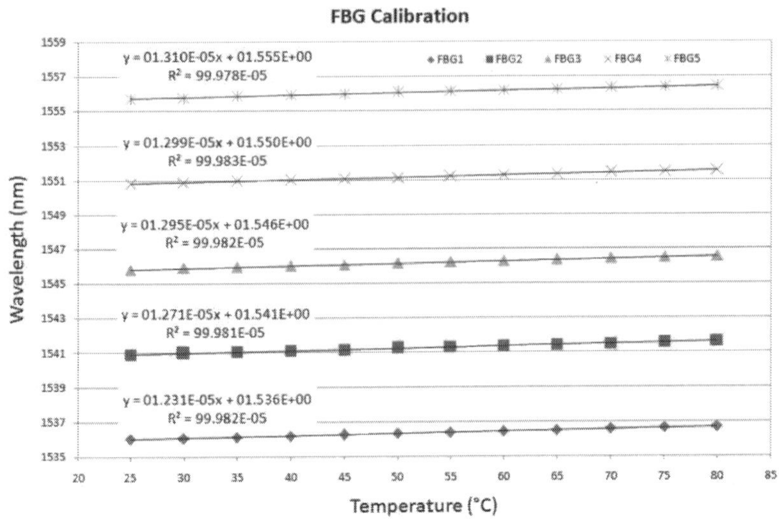

Figure 5. Wavelength shift versus temperature for each FBG [**4**].

Table 2. Calibration Parameters [**4**]

FBG #	Theoretical Sensitivity (pm/°C)	Measured Sensitivity (pm/°C)	Correlation Coefficient (R^2)	RMSE (°C)	Max. Residual Error (°C)
1	14.05	12.31	0.99982	0.00311	0.003
2	14.10	12.71	0.99981	0.00328	0.005
3	14.14	12.95	0.99982	0.00327	0.005
4	14.19	12.99	0.99993	0.00320	0.006
5	14.23	13.10	0.99978	0.00363	0.007

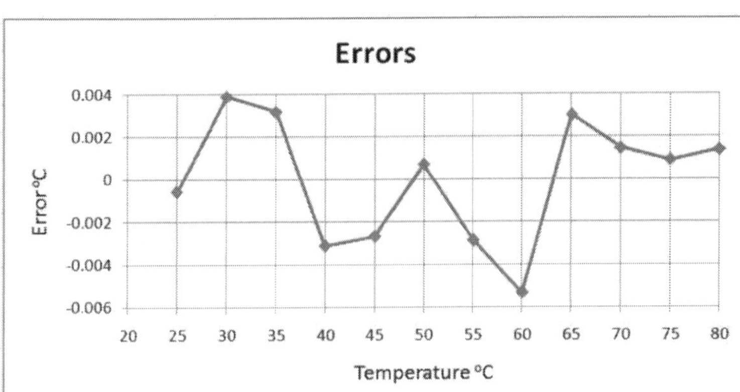

Figure 6. Error analysis for FBG 1 [**4**].

All correlation coefficients are very close to unity and errors much smaller than 1°C. These errors are a combination of the uncertainty of the interrogation system, (±1 pm) and of the thermometer used. Using the FBG's average sensitivity of 13 pm/°C (see Table 2), 1 pm in error means a temperature uncertainty of about 0.08°C which is much smaller that the error produced by the thermometer.

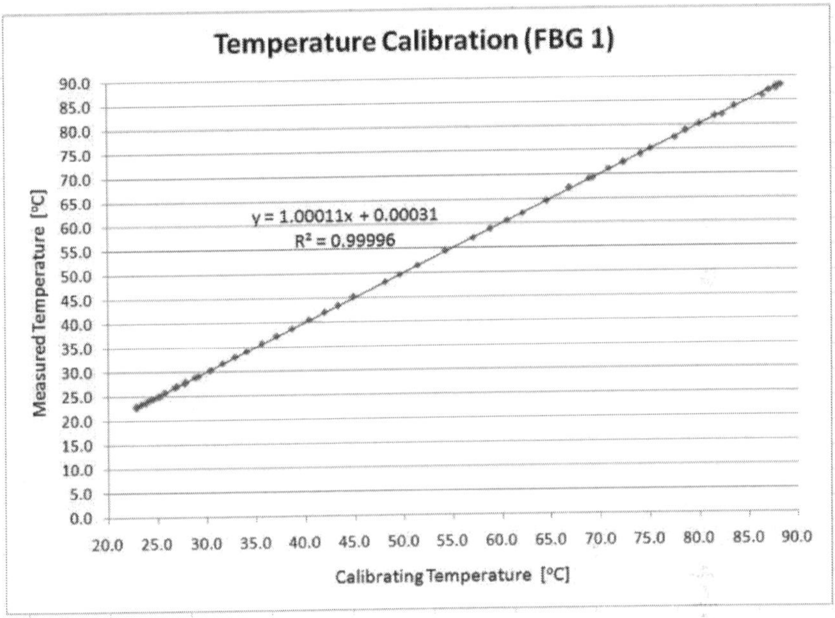

Figure 7. Temperature calibration responses of FBG 1 [**4**].

Notice in Table 2 that the theoretical sensitivities predicted by Equation (3) are different from those obtained in the calibration experiment. Also, Eq. (1) shows λ_B as a function of n_{eff}, the average index of refraction between the pristine fiber core and that of the Ultra-Violet (UV)-irradiated core. The FBG fabrication processes is not automatic and the radiation time for each FBG inscription is not the same as the laser is turned off by the operator when Bragg reflection appears above the desired level. The UV modifies the index of refraction of the fiber core and also modifies the values of η in each FBG differently. These results in the slightly dispersed sensitivities found above. This effect is confirmed by [5] that demonstrated a technique for changing the temperature responsivity of FBGs through increased UV exposure over the FBG.

From the data in Table 2 it is possible to calculate the relationship between wavelength and temperature for each FBG. to-one fitting accuracy to the third decimal place in temperature and a correlation coefficient R^2=0.99996, demonstrating a very good linearity and accuracy of FBG sensors for temperature measurements.

5. PHOTOSENSITIVITY IN OPTICAL FIBERS

Photosensitivity in glass, as mentioned in Session 2, was discovered at the Communications Research Center in Canada, in 1978 by Hill and co-workers [1]. It was a new nonlinear effect in optical fibers and was called at that time of fiber photosensitivity.

A decade later after this discovery, Meltz and co-workers [3] have proposed a model for fiber photosensitivity. What motivated their model was that at first, fiber photosensitivity was detected only in fibers containing germanium as a dopant.

The model is based on the fact that when germanium-doped silica fibers are fabricated by MCVD technique, germania (GeO_2) and silica (SiO_2) in form of gases combine in high temperatures to produce the fiber. During the process though, there is a statistical probability that products like Ge-Ge, O-Ge-O, Ge_2^0 and Ge-Si might be formed, which are defects in the fiber lattice and are called in the literature as "wrong bonds". The fiber presents strong absorption peak at 245 nm, which is associated with these defects. When these defects are irradiated with UV light some absorptions bands appear and the index of refraction increases in these points.

The origins of photosensitivity and the change of the RI as a consequence have yet to be fully understood as no single model can explain the experimental results shown in the literature. So, it becomes apparent that photosensitivity is a function of several mechanisms such as photochemical, photomechanical, thermochemical, etc. [6].

One of the models that show consistencies with experimental results seems to be the compaction model based on the Lorentz-Lorenz law which states that the RI increases with material compression. This idea was pursued

These equations were fed into the software of the FBG interrogation system which returns the temperature of each sensor in a field application. Finally, it is possible to plot the calibrating temperature against measured temperature, as shown in Figure 7, presenting a one-by [7] that used UV Laser irradiation to produce thermally reversible, linear compaction in amorphous SiO_2. An accumulated, incident dose of 2000 J/cm^2 would produce an irreversible compaction and photoetching. The above results are in accordance with the fabrication of Type I and Type IIA FBG. Also, Laser compaction results were found to be consistent with those obtained using hydrostatic pressure. Therefore it was observed an approximately linear RI versus $\Delta V/V$ agreeing with the predictions of the Lorentz-Lorenz law.

Also, in accordance with this model of densification associated with the writing of Bragg gratings, Riant and co-workers [8] observed a transition mechanism between Type I and Type II gratings.

Notice that, when a transparent material is compressed we observe two effects that interfere with the RI. One is the increase of RI due to the increase of density of the material. The other is the photoelastic effect, which is negative for many optical media, and produces an opposite effect. However the compression produces an effect much stronger than the photoelastic effect and we normally observe an increase of RI in the irradiated parts of the FBG.

With the knowledge that the UV irradiation produces an increase on the RI we can now go further and observe physically the FBG. Figure 8 shows the core's RI along the length of the fiber (z axis) with most frequent values of the parameters. The effective RI of the core is the average index of refraction of the irradiated portion of the core and is approximately 1.45. Due to the UV radiation the variation of the RI is about $\Delta n = 10^{-4}$. The grating period (Λ) is the same as the interference pattern, about 500 nm and the FBG length L_{FBG} is around 10 mm.

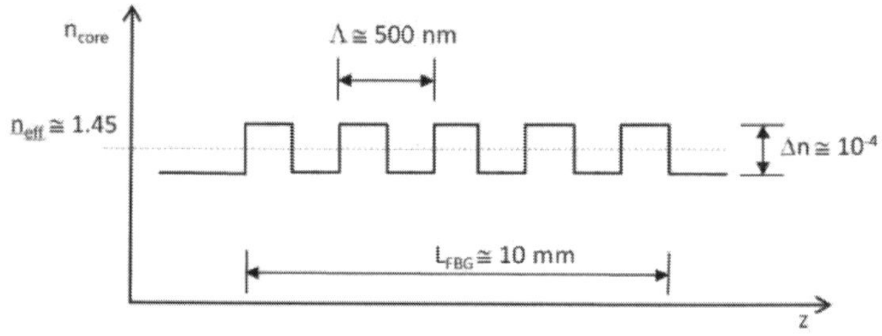

Figure 8. The Refraction Index variation of the fiber core along the length of the fiber (z axis) with most frequent values of the parameters.

The interference pattern, however, does not vary as a square wave but rather as an approximate sinusoidal waveform, which will inscribe an RI variation on the fiber of the same form, as shown in Fig. 9. In the figure, n_{eff} is the RI of the pristine fiber core, $<n_{eff}>$ is the average RI in the FBG region, Δn_{DC} is the average amount of RI increased by the UV dose and Δn_{AC} is half of the total RI variation in the FBG.

The mathematical model of the RI in the FBG area as a function of the UV radiation dose (d) in Joules and the distance *z* along the fiber's axis is [9]:

$$n_{eff}(z,d) = n_{eff} + \left(\Delta n_{DC}(d) + \Delta n_{AC}(d)\sin\frac{2\pi}{\Lambda}z\right) \tag{23}$$

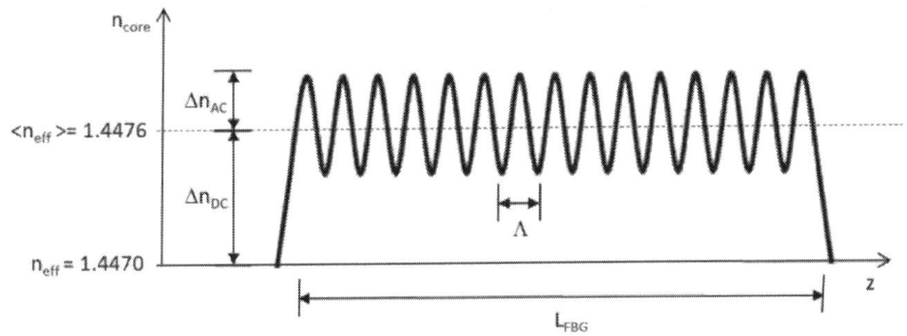

Figure 9. Variation of the Refraction Index of the fiber`s core along the length of the fiber (z axis) resultant of a sinusoidal diffraction pattern [*9*].

Notice that both Δn_{DC} and Δn_{AC} increase with UV dose and, since the UV radiation is never zero along the diffraction region, all FBG length experiences an increase of RI.

Now we can rewrite Eq. (2) using the average RI in the FBG area:

$$\lambda_B = 2\langle n_{eff}\rangle\Lambda \tag{24}$$

For an FBG inscription using the setup of Figure 11, it is necessary first to calculate the angle φ for the desired Bragg wavelength. Then, the operator monitors the reflection spectrum until the reflectivity reaches the desired value. But since $<n_{eff}>$increases during the irradiation, so does λ_B, according to Eq. (24). Figure 10 shows the progression of an FBG reflection spectrum as UV dose increases.

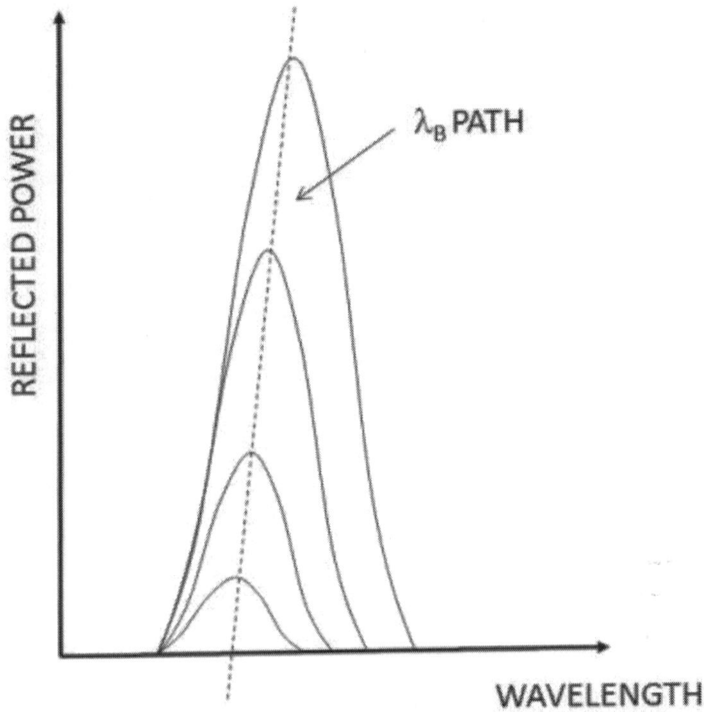

Figure 10. Progression of an FBG reflection spectrum as UV dose increases [9].

The dotted line in Figure 10 shows the path of the Bragg wavelength as UV dose increases, or same to say, the variation of $<n_{eff}>$, according to Eq. (24).

Therefore, two relationships can be obtained from Figure 10:

$$\text{Reflectivity} = \text{f(number of shots or irradiation time)} \quad (25)$$

$$\lambda B = g(\text{number of shots or irradiation time}), \quad (26)$$

where f(*) And g(*) are arbitrary functions to be determined by curve fitting.

Both reflectivity and Bragg wavelength increase with the number of laser shots because each shot represents a certain amount of UV energy and the energy is integrated producing the stress inside the fiber core. These equations will be important when an FBG is designed.

As UV dose increases, so does reflectivity, but up to a threshold above which the reflectivity starts do lower again. This is due to the competing effects between the increase of RI due to the increase of density of the material and the photoelastic effect, which is negative for many optical media. Above the threshold limit, the photoelastic effect is stronger than the increase of density, probably because this last effect saturates while the photoelastic effect does not. The threshold is about 500 mJ/cm^2 and this value is considered to be the limit separating Type I gratings to Type IIA. While during the formation of Type I FBG, λ_B experiences a red shift (see Figure 10), during the erasure of the FBG in type IIA formation, the Bragg wavelength experiences a blue-shift. Above this limit, the FBG starts to be erased until it completely disappears.

Bragg grating in germanosilacate fibers exhibits a temperature decay dependency. Type I FBGs are found to present reasonable short term stability up to 300°C, whereas Type IIA gratings exhibit very good stability up to 500°C [6].

Therefore, after fabricating a Type I FBG it is recommended to submit them to an annealing process up to a temperature that exceeds the service temperatures of the application in order to produce an accelerated ageing.

Conventional telecommunication fibers normally present around 3.5% concentration of germania doping. These fibers will weakly respond to UV radiation ($\Delta n \cong 10^{-5}$) and will grow low reflectivity FBGs. Only fiber with 5% plus GeO concentration present photosensitivity enough to be useful for FBG fabrication, but are more expensive than conventional telecommunication fibers. Those fibers with up to 30% of dopant concentration are produced by several fiber makers such as Nufern, Fibercore and IPTH.

Another dopant used is GeO co-doped with boron; these fibers present an enhanced sensitivity however causing an increase in losses. Therefore, boron co-doped fibers are not good for long sensing distance, they are limited to some few meters only in contrast with pure GeO doped fibers that can be used to remotely monitor parameters which are several kilometers away from the interrogation system.

Another way of enhancing the fiber sensitivity is by hydrogen diffusion into the fiber core. The mechanism causing an increased sensitivity is thought to be due the reaction of H_2 with GeO. In highly doped fibers there is a significant concentration of Ge-O-Ge bonds. H_2 reacts with these bonds resulting in the formation of Ge-OH, which absorbs UV radiation and therefore increasing the internal stress into the core of the fiber [10].

The hydrogen diffusion is accomplished by leaving the fiber into a tight enclosure with hydrogen at high pressure. Pressures from 20 atm to 750 atm can be used but most commonly 150 atm. Apart from increasing the fiber photosensitivity, hydrogen loading allows the fabrication of FBG in any germane silicate and germanium free fibers.

When a hydrogen loaded fiber is taken out of the high pressure vessel it is as soft as cotton string. However, by heating the fiber after the exposition, the hydrogen diffuses out in a few minutes.

Hydrogen loading can be also accomplished by a technique known as flame brushing. This technique consists in burning for 20 minutes the fiber by a hydrogen-oxygen flame reaching temperatures of 1700°C. At high temperature, the excess of hydrogen in the mixture diffuses into the fiber. The advantage of frame brushing is that it is possible to sensitize conventional telecommunications fibers. The disadvantage, however, is that the flame burns the fiber acrylate buffer in an area larger than that of the FBG itself which demands a posterior fiber recoating.

6. FABRICATIONS TECHNIQUES

As mentioned above the interest in FBG started with the possibility of inscribing the grating sideways as demonstrated by Meltz [3] and with the possibility of tuning to a desired wavelength along the telecom band. From then on, many FBG applications appeared first in telecom such as add/drop, dense wavelength division multiplexing (DWDM) mux/demux, filters, lasers, and so on. Later, with the telecommunications devices and equipments decreasing prices, FGBs started to be used as sensors in a commercial basis.

The first technique used to inscribe a FBG in the fiber was the interferometer and it is used in many different configurations (Figure 11 shows a basic interferometer). A laser beam is divided in two by a beam splitter or a prism. Each part is reflected in mirrors to meet again to form a interference pattern over the fiber to be inscribed. Cylindrical lenses concentrate the beams in the inscribing area of the fiber, about 5 mm by 200 μm in order to increase the density of the UV dose.

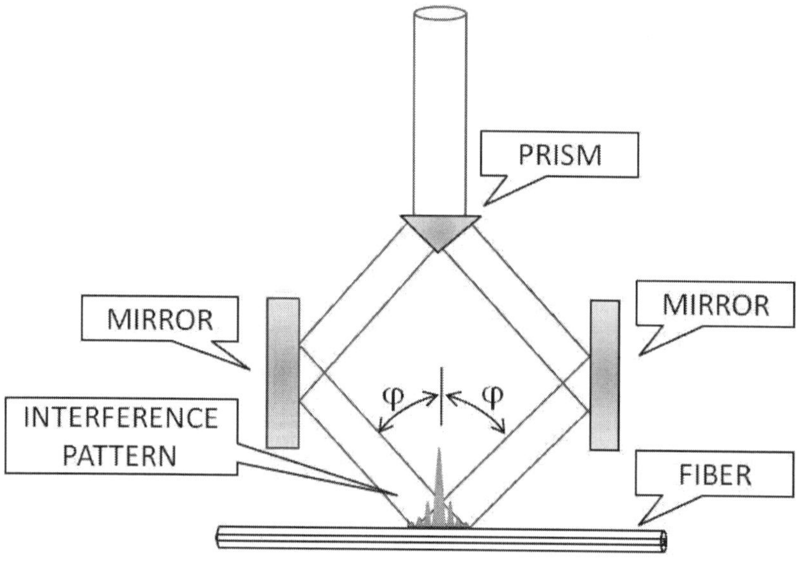

Figure 11. A basic interferometer.

The period of the interference pattern (Λ) depends on the wavelength of the light used for writing (λ_{Laser}) and also on the half-angle between the two interfering beams (φ) as shown in Eq. (27).

$$\Lambda = \frac{\lambda_{Laser}}{2\sin\varphi} \tag{27}$$

There are some disadvantages when using such arrangement as, for instance, the difficult to align the beams, the necessity to achieve a very good spatial coherence on the laser and problems associated with air flow which slightly modifies the RI of the air distorting the wave front of the beams. This effect could lead to an FBG of poor quality. The advantage is that one can adjust the mirrors in order to vary the grating period to virtually any value around the telecommunication band.

Nowadays we rely on the phase-mask technique which is a diffractive optical element that spatially modulates the UV beam with period Λ_{pm}. The phase masks are formed in a fused silica substrate by a holographic technique or electron beam lithography.

When a laser beam is incident to the phase mask a diffraction occurs and the beam is divided into several diffraction orders. The zero order is suppressed to less than 3%, but the +1 and -1 orders prevail with most of the remaining power. These two orders start from the same point on the other side of the phase mask but are divergent. At the near field an interference pattern is produced as the two orders cross each other, with a period

$$\Lambda = \frac{\Lambda_{pm}}{2} \qquad (28)$$

The optical fiber is placed in contact or in close proximity to the phase mask, inside the near field where the interference pattern is produced as shown in Figure 12. An increased power density can be achieved by the use of a cylindrical lens parallel to the fiber, before the phase mask.

The advantage of the phase mask is that its setup is much simpler because there is no need for the laser to have a good coherence and there are no mirrors to align. However, as the phase mask is such a fragile optical element, the close proximity of the fiber to the phase mask surface can scratch it. If the distance between the phase mask is increased by a few millimeters the fiber will be illuminated by a narrower interference and the FBG will be accordingly narrower. Another disadvantage of the phase mask technique is that the periodicity of the FBG inscribed is fixed by the one of the phase masks, according to Equation (28).

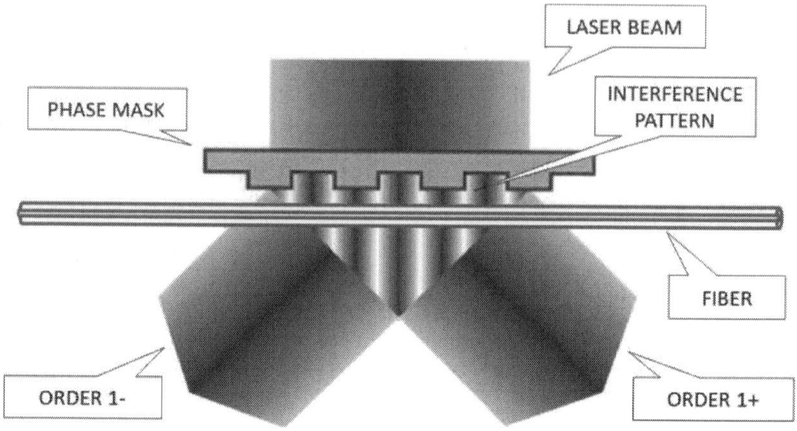

Figure 12. A laser beam inscribing gratings in a optical fiber through a phase mask.

An alternative setup is shown in Figure 13 in which the phase mask is far away from the fiber and the two mirrors redirect the +1 and -1 refracted orders back to the fiber.

The advantage of such setup is that one can adjust the Bragg wavelength by the angles of the mirrors.

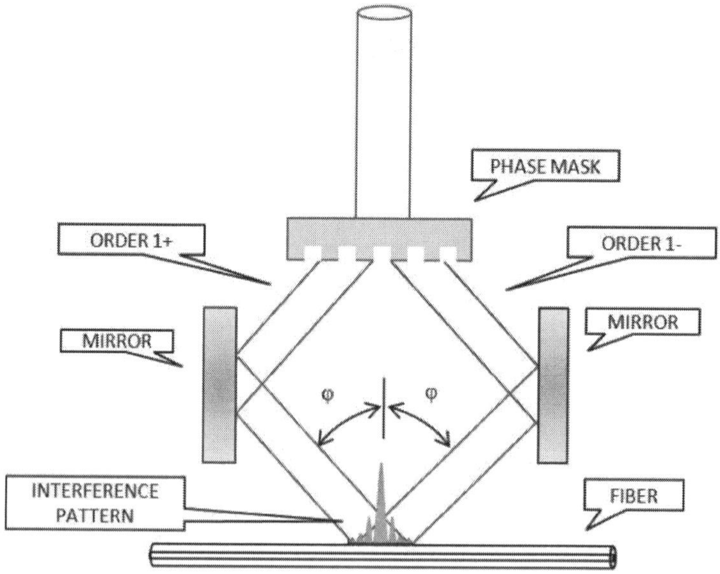

Figure 13. Most common setup used with a phase mask.

7. FBG FABRICATION PARAMETERS

When specifying an FBG the following parameters must be known:

- Central Brag wavelength, λ_B
- FBG width, FWHM
- Reflectivity

Bragg wavelength(λ_B) depends essentially on the phase mask periodicity or on the laser wavelength, and on the intersection half-angle in the case when an interferometer setup is used (Figure 11). However, the UV dose also modifies the Bragg wavelength, according to Eq. (25).

FWHM depends on the FBG length and on the UV dose, according to the following Equation [6]:

$$\mathrm{FWHM} = \lambda_B s \sqrt{\left(\frac{\Delta n}{2n_{eff}}\right)^2 + \frac{1}{N}^2} \tag{29}$$

where, Δn is the amplitude of the induced RI in the fiber, $\Delta n = 2 \times \Delta n_{AC}$ (see Figure 9), λ_B is the final Bragg wavelength, s is 1 for strong reflection grating with reflectivity close to 100%, or 0.5 for weak gratings, N is the number of grating planes, $N = L_{FBG}/\Lambda$.

Reflectivity, as it has been seen in later sessions, is a function of the UV dose in J/m^2, the amount of Germania doping in the fiber and the hydrogenation processes. Equation (26) can be useful for predicting the reflectivity and this parameter can also be adjusted by varying the FBG length. This is accomplished by adjusting the laser beam width. One can produce lengths as small as 2.5 mm obtaining, thus very low reflectivity and lengths as large as 15 mm obtaining a reflectivity around 100%.

Therefore, when projecting an FBG inscription, one has to carry out an inverse engineering to preview how much λ_B will displace during the inscription to the desired reflectivity and decrease this value to the desired Bragg wavelength in order to adjust the mirrors accordingly.

As none of the above parameters are not precisely known, the best way to know the exact time of irradiation is by experimental tests.

8. INTERROGATION TECHNIQUES FOR FBG SENSORS

The main challenge when it comes to FBG sensing is the method to demodulate its wavelength changes. The use of FBG sensors is connected to the development of techniques to interrogate these sensors and detect Bragg wavelengths shifts as a function of the parameter being measured.

The easiest way to interrogate an FBG is by the use of an optical spectrum analyzer (OSA) which performs a direct measurement of the reflection spectrum of the FBG. The other method is based on the conversion of wavelength variations into optical power intensities.

The technique using an optical spectrometer is very simple. The interrogation system consists of a broadband optical source which illuminates the FBGs. Their reflected peaks, which are represented by each Bragg wavelength λ_B, are directed to the OSA and monitored by a computer, as shown in Figure 3.

Although being a simple demodulation technique, it presents some disadvantages: first, the commercial OSAs are heavy and expensive equipment, besides being inappropriate for field applications. Moreover, most of the spectrum analyzers are limited to static measurements, so they don't have sufficient resolution concerning the response time when a number sensors are being interrogated. A conventional OSA will present an accuracy of about ±50 pm, which would produce errors of about ±4°C and ±60 με, according to equations (9) and (10). For most applications these errors are unacceptable if compared to conventional resistive temperature detectors (RTD) and strain gauges. To detect small variations in wavelength the development of new techniques must also ensure essential characteristics such as static and dynamic measurements, real-time measurements, accuracy, resolution, and low cost, all necessary conditions for field applications. In conclusion this technique is only useful for laboratory applications and tests.

The setup shown in Figure 3 is also commercially available by a few companies as standalone equipment and appropriated to go to the field. In this case they feature resolutions as good as ±2 pm and can be programmed to monitor specific parameters as pressure, strain, temperature, etc. However, due to the high cost, they will be applicable to solve monitoring needs of industries only if the project included as many sensors as possible to be monitored by one single unit to have the total price divided by the number of sensors.

The other demodulation technique uses a Fabry-Perot (FP) tunable filter. Although the interferometric FP filter method is a consolidated technology, showing high resolution and accuracy, it still presents a moderate cost. The tunable optical filter scheme is based on a Fabry-Perot extrinsic cavity, which is adjusted by mirrors and by varying the internal cavity of a PZT crystal by means of an external power supply, enabling the filter adjustment and selection of the desired wavelength. This relationship between the changes in the filter wavelength as a function of an applied voltage is linear. Defects in the geometry of the lens during the filter manufacturing process can cause instability in the measurement system, so that the optical spectrum of the filter is not entirely symmetrical.

The demodulation setup using a FP filter is shown in Figure 14. A broadband light source was used to illuminate the FBG sensor via an optical circulator. The reflected spectrum of the sensor passes through the FP tunable filter with a 0.89 nm bandwidth.

This demodulation technique is based on the same principle of an FM radio signal demodulated by an edge filter. The signal waveband is made to vary at the wedge of a filter that will transmit a variable power proportional to the variation of the signal frequency. In our case, proportional to Bragg shift. In reality, the transmitted signal through the filter is proportional to the convolution between the signal power and the filter response.

The optimum position of the center wavelength of the FP filter is chosen by an algorithm described by [11]. The dashed area on the spectra drawing (inset in Figure 14) is the intersection between the spectrum of the reflected signal and the band pass of the FP filter. The integral of this area represents the total light power that reaches the photodetector.

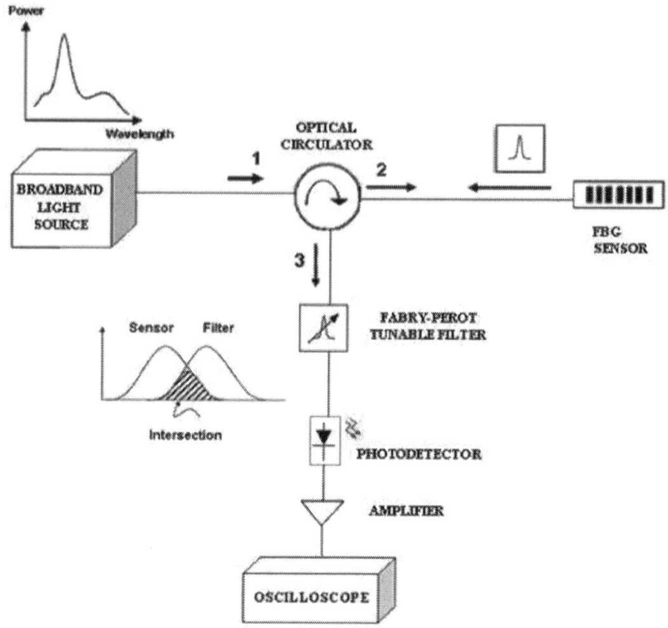

Figure 14. The interrogation setup using a Fabry-Perot filter (adapted from [**11**]).

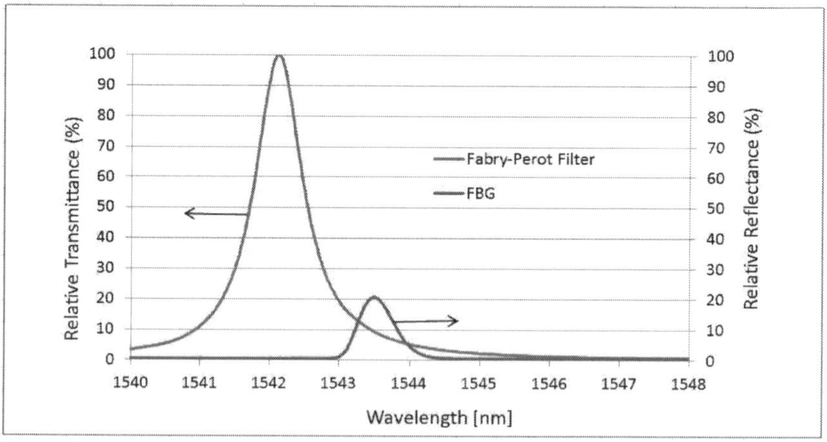

Figure 15. Spectral curves for the Fabry-Perot filter, $F_{FP}(\lambda)$, and the FBG, $F_{FBG}(\lambda)$.

The spectral curves for the FP filter, FFP(λ), and for the FBG, $F_{FBG}(\lambda)$, are shown in Figure 15, where the sensor is at quiescent state. The vertical axes show the relative transmittance of the FP filter and the relative reflectance of the FBG sensor, respectively. The numerical convolution $F_{FP}(\lambda)*F_{FBG}(\lambda)$ represents the available power to the photodetector as a function of the wavelength shift. The convolution curve is shown in Figure 16.

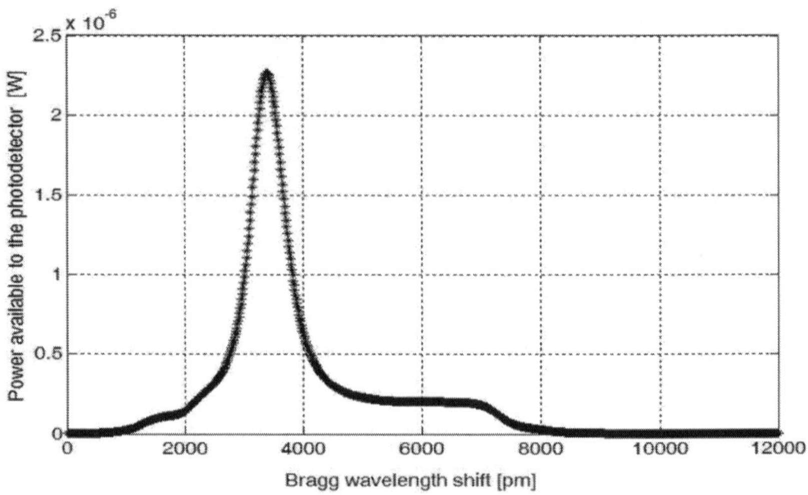

Figure 16. The convolution between $F_{FP}(\lambda)*F_{FBG}(\lambda)$.

Instead of an FB filter, it is possible to use another FBG, in this case used as a dichroic mirror, differently from the FP filter which acts by light transmission. A broadband light source injects light into port 1 of the optical circulator 1. The light circulates to port 2, illuminating the FBG sensor. The reflection spectrum of the FBG sensor is deviated to port 3, and enters through port 1 of circulator 2. Circulator 2 deviates the signal to the twin FBG filter, through port 2. Only the superimposed wavelengths (inset graphic) reflect back to circulator 2, which deviates the light to the photodetector through port 3.

This demodulation scheme is very simple, and reduces the cost of the setup implementation; however, the twin FBG must be manufactured at an exact wavelength to provide an optimized operation procedure.

REFERENCES

1. K. O. Hill, Y. Fujii, D. C. Johnson, B. S. Kawasaki, Photosensitivity in optical fiber waveguides: application to reflection fiber fabrication", Appl. Phys. Lett. 32 (10): 647, 1978

2. K. O. Hill, Photosensitivity in Optical Fiber Waveguides: From Discovery to CommercializationIEEE Journal on Selected Topics in Quantum Electronics, 6611861189November/December 2000

3. G. Meltz, W. W. Morey, and W. H. Glenn, Formation of Bragg gratings in optical fibers by a transverse holographic method"Opt. Lett. 14 (15): 823 EOF5 EOF1989

4. M. M. Werneck, R. C. Allil, and B. A. Ribeiro, Calibration and Operation of a Fiber Bragg Grating Temperature Sensing System in a Grid-Connected Hydrogenerator", IET Science, Measurement & Technology, accepted to publication on September, 2012

5. B. A. L Gwandu,. and W Zhang, , W., "Tailoring the temperature responsivity of fibre Bragg gratings", Proceedings of IEEESensors,

6. A. Othonos, K. Kalli, Fiber Bragg Gratings- Fundamentals and Applications in Telecommunications and Sensing", Artech House, 1999

7. C. Fioria, and R. A. B. Devinea, Ultraviolet Irradiation Induced Compaction and Photoetching in Amorphous Thermal SiO2". MRS Proceedings of the Fall Meeting, 611985

8. I. Riant, S. Borne, P. Sansonetti, B. Poumellec, Evidence of densification in UV-written Bragg gratings in fibres", in Photosensitivity and Quadratic Nonlinearity in Glass Waveguides: Fundamentals and Applications, 5255Postconference Edition, Optical Society of America, 1995

9. F. Jülich, and J. Roths, Determination of the Effective Refractive Index of Various Single Mode Fibres for Fibre Bragg", Proceedings of SENSOR+TEST Conference (OPTO-2009Nürnberg, 119124

10. P. J. Lemaire, M. Hill, and T. Erdogan, Hydrogen-enhanced UV photosensitivity of optical fibers: Mechanisms and reliabilityin Photosensitivity and Quadratic Nonlinearity in Glass Waveguides

Fundamentals and Applications, September 9-11, 1995Portland, Oregon, Technical Digest Series, 227881

11. B. A. Ribeiro, M. M. Werneck, and J. L. Silva-neto, A Novel Optimization Algorithm to Demolutate a PZT-FBG sensor in AC High Voltage Measurements", in review at IEEE Sensors Journal, September, 2012

CHAPTER 4

Review of Turn around Point Long Period Fiber Gratings

Monika Gambhir, Shilpi Gupta

Department of Electronics & Communication Engineering, SVNIT, Surat, India

ABSTRACT

Long period fiber gratings are emerging as a potential candidate in the list of surrounding refractive index optical fiber sensors. Their sensitivity can be enhanced greatly if the grating period, fiber dimensions and surrounding refractive index are optimized in a way to operate at a point called turn around point on phase matching curves of these gratings. Turn around point LPFGs are well known for their ultrahigh sensitivity to external parameters. Potential of operating LPFG at or near turn around point has been investigated by many researchers in various applications including physical parameter sensing, adulteration detection, radiation dose, etc. Since TAP LPFGs are in investigation phase therefore a lot of rigorous & efficient work in finding techniques for optimizing their potential as sensor in chemical, biochemical, structural health monitoring is still to be carried out. A brief review of work carried out in this domain till now is presented here and key findings from literature review are highlighted.

KEYWORDS

LPFG (Long Period Fiber Grating), TAP (Turn around Point), SRI (Surrounding Refractive Index), LP (Linearly Polarized)

1. INTRODUCTION

Long period fiber gratings are passive optical fiber sensors emerged in recent years. There are mainly two types of Index (refractive index) gratings in fiber optics: FBG (Fiber Bragg Gratings) & LPFG (Long Period Fiber Gratings). Structurally, both types of gratings are identical except that the periods of Fiber Bragg gratings are two orders of magnitude smaller than those of LPGs. FBGs induce contra directional coupling. Light energy from forward propagating modes is transferred to modes traveling in the opposite direction. FBGs are often referred to as short period, reflection gratings because of this effect. LPFGs, on the other hand, couple energy co-directionally. Energy from the forward propagating incident modes is transferred to either forward propagating cladding or radiation modes depending on the structure of the waveguide [1] -[6] .

Fiber Bragg gratings have already proved their excellence in physical parameter sensing, e.g. temperature, strain, torsion, bend, etc. and in many practical applications including structural health monitoring which is required for existing smart world era [7] . LPFGs are known to have high refractive index sensitivity in comparison to its counterpart Fiber Bragg Gratings. The evanescent field of coupled mode that extends more into the surroundings in LPFGs enables it to be used as a significant surrounding refractive index sensor.

Sensitivity of these gratings greatly depends upon the choice of grating period. Dispersion curves drawn between resonance wavelength (λ_{res}) and grating period (Λ) for higher order cladding modes indicate the presence of turn around point where the slope of the curve changes sign from positive to negative. LPFGs fabricated at these turn around points appear to be ultrahigh sensitive as the slope $\frac{d\lambda_{res}}{d\Lambda}$ of the dispersion curve is infinite.

For lower order modes these turn around points occur at higher wavelengths outside the optical communication window. The possibility of interaction of evanescent field with surroundings improves with increase in order of the cladding mode being coupled to the fundamental mode due to larger coupling coefficients [8] .

In this paper, systematic review of the LPFGs operating at turn around point is presented. Section 2 highlights the presence and significance of turn around points that exist on the characteristic curves of cladding modes. Fabrication techniques, methods used for tuning of turn around point have also been discussed in this section. Reported applications of TAP LPFGs have been summarized in Section 3. Section 4 provides an overview of thin film coated LPFGs and recently reported applications of thin film coated LPFGs operating at turn around points. The paper is concluded in Section 5.

2. ULTRAHIGH SENSITIVE TAP LPFGS

Periodic modulation of refractive index in optical fiber causes coupling of fundamental core mode with either co-propagating or counter propagating cladding modes. Coupling of co-propagating fundamental guided mode LP_{01} and cladding modes represented by LP_{0m} takes place in long period gratings according to the phase matching condition, which results in series of attenuated resonance peaks in transmission spectrum. The phase matching condition is given by

$$\beta_{01} - \beta_{cl,m} = \frac{2\pi}{\Lambda} \tag{1}$$

where β_{01} is propagation constant of the fundamental mode, $\beta_{cl.m}$ is propagation constant of mth cladding mode and Λ is period of grating. Long period fiber grating sensor configuration shown in Figure 1 represents the presence of evanescent field of coupled modes that can interact with surroundings.

Mathematical expressions derived in [8] for calculating shift in resonance wavelength with respect to change in surrounding refractive index is given by

$$\frac{d\lambda_{res}}{dn_{surr}} = \lambda_{res} \cdot \gamma \cdot \Gamma_{res} \tag{2}$$

where λ_{res} is resonance wavelength, n_{surr} is surrounding refractive index, γ is general waveguide dispersion and Γ_{res} represents surrounding refractive index dependence on waveguide dispersion respectively. Slopes of the phase matching curves drawn for B/Ge co-doped fiber, with cladding modes of order m = 8 to 20 change their sign from positive to negative at points called turn around points as indicated in Figure 2. These points offer ultrahigh sensitivity because a change in wavelength corresponding to change in grating period is infinite. Grating period may change due to change in effective refractive indices of modes which also depends on surrounding refractive index. Therefore any small change in surrounding refractive index may cause significant shift in resonance wavelength. If Λ is short, the phase matching curves corresponding to coupling between core and higher order cladding modes are parabolas over the wavelength range of 800 - 2000 nm [8] -[11] .

Phase matching curves plotted between grating period and wavelength are essential in analysis of long period fiber gratings. Existence of turnaround point in the phase matching curves was revealed for first time in [8] . Coupling to higher order cladding modes increases sensitivity to surrounding refractive index because of extension of evanescent field in surrounding with increase in cladding mode order. Moreover coupling strength of fundamental mode and odd order cladding modes is better as compared to that with even order cladding modes. This is because odd order cladding modes have a peak of intensity that lies in the core. Transmission spectra of Long period fiber gratings designed to operate at TAP indicate that any change in SRI causes change in peak attenuation at resonant wavelength which opens the possibility of simple & easy amplitude based demodulation technique. Further, if the SRI is more than refractive index of air i.e. if LPFG is immersed in a liquid, there are two resonant wavelengths in transmission spectrum and change in SRI causes shift in these two resonant wavelengths & therefore wavelength based demodulation can be used at the expense of decrease in sensitivity.

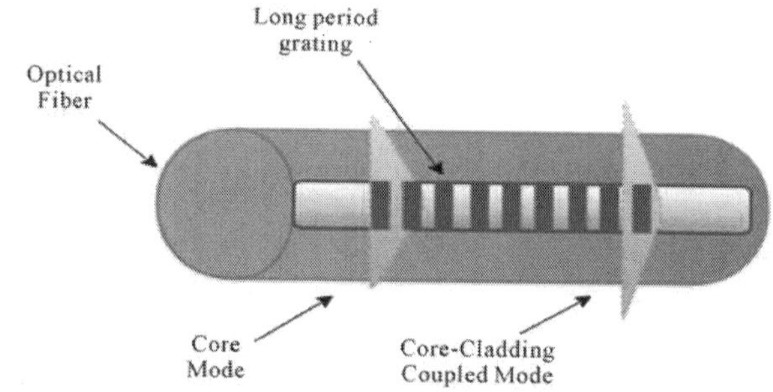

Figure 1. Long period fiber grating refractive index sensor.

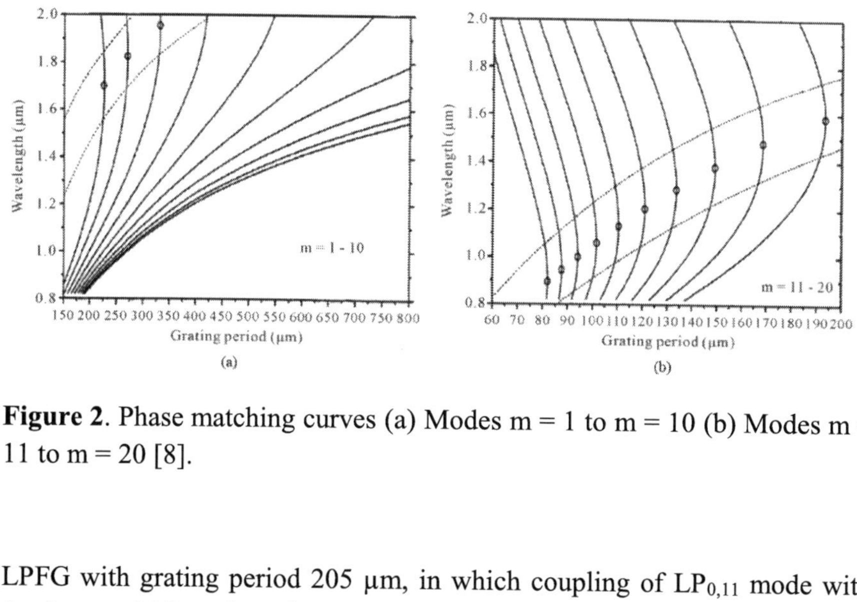

Figure 2. Phase matching curves (a) Modes m = 1 to m = 10 (b) Modes m = 11 to m = 20 [8].

LPFG with grating period 205 μm, in which coupling of $LP_{0,11}$ mode with fundamental LP_{01} takes place exactly at turn around point, is reported in [12] . Designed grating is used for fuel adulteration detection. Single attenuation peak splits into two when grating is immersed in petrol having refractive index greater than 1.0 which is in line with [8] . Reducing cladding diameter of LPFG operated near TAP greatly enhances the sensitivity [13] . LPFG with a grating period of 165 μm in SMF-28e fabricated which results in coupling of fundamental mode with $LP_{0,12}$ cladding mode. Figure 3 depicts transmission spectra of LPFG with SRI 1.0 and 1.333, after reducing the cladding diameter to different values [14] . It may be observed that transmission depth decreases with the increase in reduction of the cladding diameter and single attenuation peak splits into two when SRI is more than 1.

Grating period of less than 250 μm is required to enable coupling to higher order cladding modes. Therefore sophisticated techniques with high precision are required for fabrication of fiber gratings at turnaround points. CO_2 laser scanning [12] [15] - [19] , KrF excimer Laser [8] [14] [20] - [23] & Femtosecond laser [24] - [26] fabrication techniques have been used for fabrication of turn around or near turn around point long period fiber grating fabrication in different types of fibers. Recently TAP LPFG has been fabricated using KrF excimer laser in [9] . Gratings fabricated by CO_2 scanning are thermally stable & require no post processing technique in long period gratings. Operating LPFG exactly at TAP requires high precision in fabrication, a slight change in effective refractive indices of modes cause turn around point to shift. Exposure to UV [27] , Gamma radiation [16] , taper tuning [16] , etching of cladding [14] , overlay coating [28] is reported for tuning of long period grating at turn around point. Investigation of transmission spectra of LPFG is also carried out in photonic crystal fibers [25] [26] apart from standard single mode & photosensitive fibers.

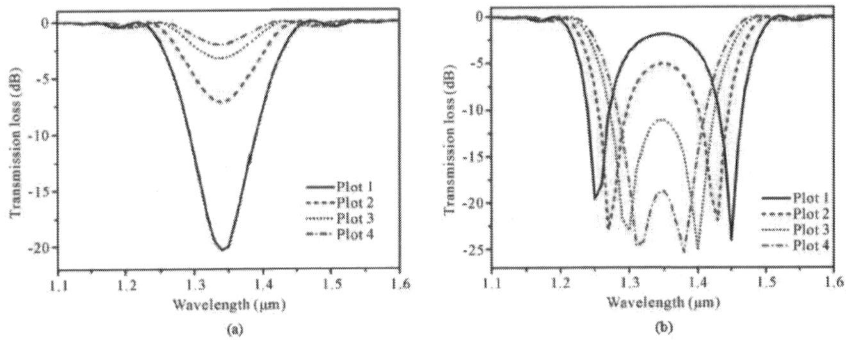

Figure 3. (a) Transmission Spectra of cladding ethed LPFG in air (b) Transmission Spectra of cladding ethed LPFG in water (Plot 1: TAP 20 dB, Plot 2: TAP 7 dB, Plot 3: TAP 3 dB, Plot 4: TAP 2 dB) [14] .

3. REFRACTOMETRIC SENSORS BASED ON TAP LPFG

Propagation characteristics of modes in optical fiber with long period fiber gratings are strong function of the refractive index of surrounding medium. These refractometric sensors have attracted attention in the field of food quality control, structural health monitoring and biomedical applications recently.

Performance of Turnaround point LPFGs have been investigated in a variety of applications including temperature, strain & refractive index sensitivity. Errors in refractive index sensing may result due to temperature variations that needs to be considered. Temperature compensated TAP LPFG have also been reported in [23] .

Fuel adulteration detection results using non TAP LPFG [20] & TAP LPFG [12] have been presented to depict the ultrahigh sensitivity of TAP LPFGs. Transmission spectra shown in Figure 4 & Figure 5 indicate that sensitivity of non TAP LPFG is 0.6 nm/% & that of TAP LPFG is 0.949 nm per percentage change in concentration of kerosene in petrol from 0 to 10%. Slight improvement in sensitivity is obtained by reducing cladding diameter as indicated in Figure 4(b).

Ultra high sensitivity of turn around point long period gratings can be used for improving the adulteration detection of the edible food products like oils, milk. Appreciably high sensitivity of turnaround point long period gratings in different applications have been summarized in Table 1.

4. THIN FILM COATED TAP LPFGS

Sensitivity of long period gratings can be improved by coating thin films of nm thickness and having refractive index greater than that of silica. Figure 6 shows four layer model used for analysis of thin film coated LPFGs with core, cladding, overlay and surroundings as each layer of the structure having refractive indices n_1, n_2, n_3 and n_4 respectively. Long period gratings offer appreciable shift in wavelength towards longer wavelengths when the SRI is less than that of the cladding & is attributed to total internal reflection.

When SRI is more than that of cladding, shift is towards shorter wavelengths and is quite small to be considered for demodulation. The reduced sensitivity of LPFG when SRI is greater than cladding is attributed to weak Fresnel's reflections. This can be enhanced by nm thick layers of thin films of high refractive index materials, e.g. tricosenoic acid, titanium dioxide, PDDA polymer, etc. [29] - [34] .

Coating of thin film material on long period fiber grating increases effective refractive indices of the cladding modes. After certain thickness of the film, cladding modes become guided in the overlay hence evanescent field of guided modes having significant energy is available near the surrounding for interaction. The reorganization of modes depends on overlay material thickness & refractive index and also on surrounding refractive index. The transition of modes occurs faster for higher refractive index overlays [34] - [36] . Transmission spectra of thin films of various refractive indices & thickness on LPFG are analyzed in [30] .

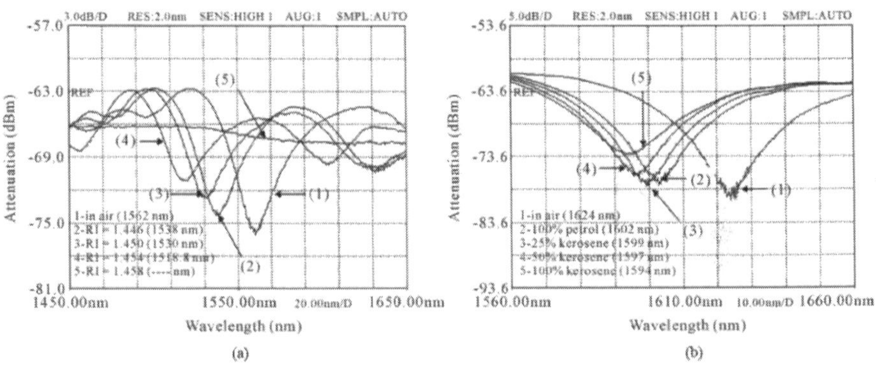

Figure 4. (a) Transmission Spectra of LPFG with different Kerosene in petrol (b) Transmission Spectra of ethed LPFG with different percentage of Kerosene in petrol [20] .

Temperature compensated chemical sensor using silica nano particles as overlay is reported in [37] . Coating is done on a pair of cascaded turnaround point long period fiber gratings. Detection of very low concentration Escherichia coli bacteria with ultra high sensitivity of 2321 nm/RIU is reported in [38] . In [39] Al_2O_3 is used as nano film material and transition region found to be in range 1.4524 - 1.461 with 3000 nm/RIU for 100 nm thickness of overlay which might not be possible with uncoated LPFG as shown in Figure 7(a). With further increase in thickness, transition region shifts to lower refractive index range. Long period gratings operating at TAP and coated with thin films of high refractive index materials have been reported in [34] [37] - [42] for chemical/biochemical sensing applications.

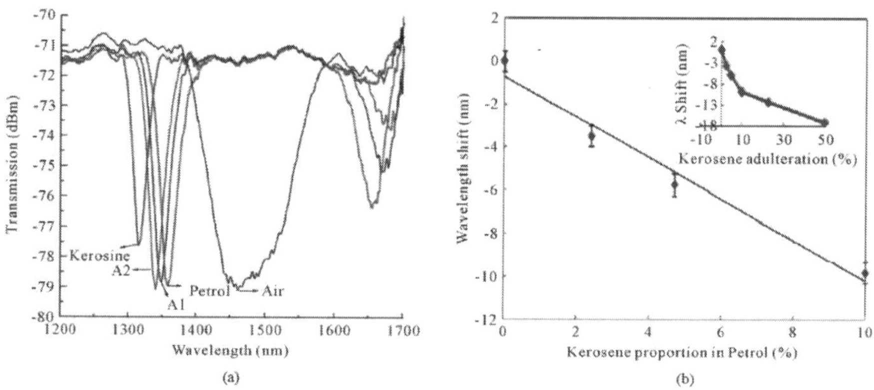

Figure 5. (a) Transmission Spectra of TAP LPFG with different percentage of Kerosene in petrol A1: 10%, A2: 50 % (b) Shift in TAP LPFG resonance peak as a function of Kerosene in petrol [12] .

Table 1. Reported TAP LPFGs for various applications.

Type of Fiber	Grating Period	TAP/Near TAP	Cladding Mode	Detection Range	Application	Sensitivity
B/Ge co-doped SMF	208	TAP	$LP_{0,11}$	1.397 - 1.432	Fuel Adulteration	0.949 nm/% change in Kerosene in Petrol
SMF-28e	165	TAP	$LP_{0,12}$	1.335 - 1.360	SRI Sensitivity	1847 nm/RIU
B/Ge codoped SMF	206	Near TAP	$LP_{0,11}$	6 - 65 KGy	Gamma radiation dose	80 nm for 65 KGy & 35 nm for 6 KGy
SMF-28	230.89	TAP	Not Specified	1.333 - 1.353	Tempaerature, Strain & SRI Sensitvity	1.63 nm/°C, 37 pm/µε & 2500 nm/RIU
B/Ge co-doped SMF	180	TAP	$LP_{0,12}$	325 - 1300 µε	Strain	1.77 dB/µε
B/Ge codoped SMF	206	Near TAP	$LP_{0,11}$	0 - 1300 µε	Strain	1.33 dB/µε

TAP LPFG coated with $(PDDA/SiO_2)_{10}$ film has been investigated for sensitivity to Ammonia in water. Transmission spectra with different concentrations of Ammonia indicated in Figure 7(b). The minimum detection level is found to be 0.14 ppm & response time of sensor is approximately 100 seconds [34] . Filtering characteristics of single and multiple, π phase shifted, thin film coated turn around point LPFGs have been simulated and analyzed in [43] . Parameters obtained need to be optimized for a particular application.

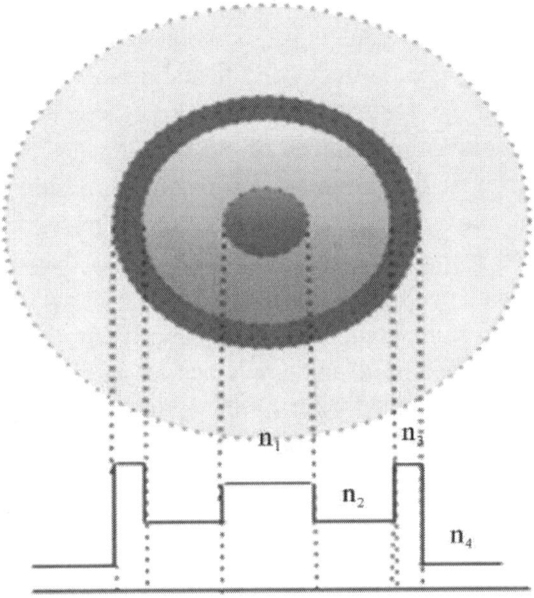

Figure 6. Four layer geometry for thin film coated LPFGs.

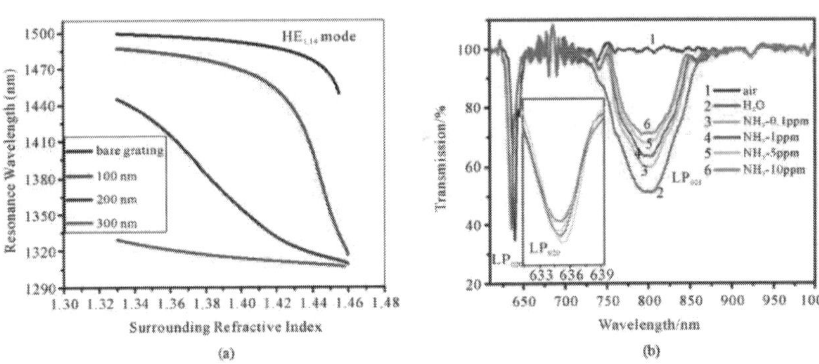

Figure 7. Response of resonance wavelength for different Al_2O_3 nano film thickness [39] (b) Transmission spectra of thin film coated PMTP LPFG for different Ammonia Concentration in water [34] .

5. CONCLUSION

SRI sensitivity improvement of TAP LPFGs can be observed in terms of significant wavelength shift & amplitude variations of the attenuation bands enabling use of wavelength and amplitude based on demodulation techniques respectively. Most of the SRI sensors based on long period operating at turn around points have been investigated in applications where the SRI is less than that of cladding. To operate these gratings as sensors in applications with SRI greater than that of cladding, coating of high refractive index materials can be done, which causes reorganization of the cladding modes and enhances the possibility of surrounding region to interact with evanescent field of cladding modes. Efficient optimization techniques for overlay material thickness and its refractive index for tuning the transition region in desired refractive index range are required to utilize LPFG in commercial applications. Further variation in physical parameters may result in false readings so some compensation techniques for these parameters are also required to be analyzed and implemented carefully. Performance of TAP LPFGs, thin film coated TAP LPFGs and a combination of these can be investigated in medical diagnostics, food quality control and biochemical hazard detection.

REFERENCES

1. Fidanboylu, K. and Efendioglu, H.S. (2009) Fiber Optic Sensors and Their Applications. 5th International Advanced Technologies Symposium (IATS'09), Karabuk, 13-15 May 2009, 1-6.
2. Hill, K.O. and Meltz, G. (1997) Fiber Bragg Grating Technology Fundamentals and Overview. Journal of Lightwave Technology, 15, 1263-1276.
3. Miguel, J., Higuera, L. and Cobo, L.R. (2015) Fiber Optic Sensors in Structural Health Monitoring. Journal of Light- wave Technology, 29, 150-164.
4. James, S.W. and Tatam, R.P. (2003) Optical Fibre Long-Period Grating Sensors: Characteristics and Application. Measurement Science Technology, 14, R49-R61.
5. Patrick, H.J., Kersey, A.D. and Bucholtz, F. (1998) Analysis of the Response of Long period Fibre Gratings to External Index of Refraction. Journal of Lightwave Technology, 16, 1606-1612.

6. Akki, J.F., Lalasangi, A.S., Raika, P.U., Srinivas, T., Laxmeshwar, L.S. and Raikar, U.S. (2013) Core-Cladding Mode Resonances of Long Period Fiber Grating in Concentration Sensor. Journal of Applied Physics, 4, 41-46.
7. Taghipour, A., Rostami, A., Bahrami, M., Baghban, H. and Dolatyari, M. (2014) Comparative Study between LPFG- and FBG-Based Bending Sensors. Optics Communications, 312, 99-105.
8. Shu, X.W., Zhang, L. and Bennion, I. (2002) Sensitivity Characteristics of Long-Period Fiber Gratings. Journal of Lightwave Technology, 20, 255-266.
9. Besley, J.A., Wang, T. and Reekie, L. (2003) Fiber Cladding Mode Sensitivity Characterization for long-Period Gratings. Journal of Lightwave Technology, 21, 848-853.
10. Hochreiner, H., Cada, M. and Wentzell, P.D. (2008) Modeling the Response of a Long-Period Fiber Grating to Ambient Refractive Index Change in Chemical Sensing Applications. Journal of Lightwave Technology, 26, 1986-1992.
11. Koyamada, Y. (2001) Numerical Analysis of Core-Mode to Radiation-Mode Coupling in Long-Period Fiber Gratings. IEEE Photonics Technology Letters, 13, 308-310.
12. Kher, S., Chaubey, S., Kishore, J. and Oak, S.M. (2013) Detection of Fuel Adulteration with High Sensitivity Using Turnaround Point Long Period Fiber Gratings in B/Ge Doped Fibers. IEEE Sensors Journal, 13, 4482-4486.
13. Kashyap, R. (2009) Fiber Bragg Gratings. Elsevier.
14. Biswas, P., Basumallick, N., Bandyopadhyay, S., Dasgupta, K., Ghosh, A. and Bandyopadhyay, S. (2015) Sensitivity Enhancement of Turn-Around-Point Long Period Gratings by Tuning Initial Coupling Condition. IEEE Sensors Journal, 15, 1240-1245.
15. Chaubey, S., Kher, S., Kishore, J. and Oak, S.M. (2014) CO2 Laser-Inscribed Low-Cost, Shortest-Period Long-Period Fiber Grating in B-Ge Co-Doped Fiber for High-Sensitivity Strain Measurement. Pramana, 82, 373-377.
16. Chaubey, S., Kher, S. and Oak, S.M. (2011) Radiation and Taper Tuning of Long Period Grating for High Sensitivity Strain Measurement. 7th Workshop on Fiber and Optical Passive Components (WFOPC), Montreal, 13-15 July 2011, 1-4.
17. Lan, X., Han, Q., Huang, J., Wang, H., Gao, Z., Kaur, A. and Xiao, H. (2013) Turn-Around Point Long-Period Fiber Grating Fabricated by

CO2 Laser for Refractive Index Sensing. Sensors and Actuators B: Chemical, 177, 1149-1155.
18. Lan, X., Han, Q., Wei, T., Huang, J. and Xiao, H. (2011) Turn-Around Point Long-Period Fiber Gratings Fabricated by CO2 Laser Point-by-Point Irradiations. IEEE Photonics Technology Letters, 23, 1664-1666.
19. Kher, S., Chaubey, S., Kashyap, R. and Oak, S.M. (2012) Turn-Around Point Long-Period Fiber Gratings (TAP-LPGs) as High-Radiation-Dose Sensors. IEEE Photonics Technology Letters, 24, 742-744.
20. Mishra, V., Jain, S.C., Singh, N., Poddar, G.C. and Kapur, P. (2008) Fuel Adulteration Detection Using Long Period Fiber Grating Sensor Technology. Indian Journal of Pure & Applied Physics, 46, 106-110.
21. Iadicicco, A., Paladino, D., Pilla, P., Campopiano, S., Cutolo, A. and Cusano, A. (2012) Ch. 14: Long Period Gratings in New Generation Optical Fibers. In: Yasin, M., Harun, S.W. and Arof, H., Eds., Recent Progress in Optical Fiber Research, InTech, 291-326.
22. Chiavaioli, F., Biswas, P., Trono, C., Bandyopadhyay, S., Giannetti, A., Tombelli, S., Basumallick, N., Dasgupta, K. and Baldini, F. (2014) Towards Sensitive Label-Free Immunosensing by Means of Turn-Around Point Long Period Fiber Gratings. Biosensors and Bioelectronics, 60, 305-310.
23. Garg, R., Tripathi, S.M., Thyagarajan, K. and Bock, W.J. (2013) Long Period Fiber Grating Based Temperature- Compensated High Performance Sensor for Bio-Chemical Sensing Applications. Sensors and Actuators B: Chemical, 176, 1121-1127.
24. Allsop, T., Kalli, K., Zhou, K., Lai, Y., Smith, G., Dubov, M., Webb, D.J. and Bennion, I. (2008) Long Period Gratings Written into a Photonic Crystal Fiber by a Femtosecond Laser as Directional Bend Sensors. Optics Communications, 281, 5092-5096.
25. Kanka, J. (2012) Design of Turn-Around-Point Long-Period Gratings in Ge-Doped Photonic Crystal Fibers for Evanescent Sensing. Proceedings of SPIE, 8426, 34-40.
26. Kanka, J. (2013) Design of Turn-Around-Point Long-Period Gratings in a Photonic Crystal Fiber for Refractometry of Gases. Sensors and Actuators B: Chemical, 182, 16-24.
27. Shu, X., Zhu, X., Wang, Q., Jiang, S., Shi, W. and Huang, Z. (1999) Dual Resonant Peaks of LP Cladding Mode in Long-Period Gratings. Electronics Letters, 35, 649-651.
28. Smietana, M., Koba, M., Mikulic, P. and Bock, W.J. (2014) Tuning Properties of Long Period Gratings by Plasma Post Processing of Their

Diamond Like Carbon Nano-Overlays. Measurememt Science and Technology, 25, Article ID: 114001.

29. Ishaq, I.M., Quintela, A., James, S.W., Ashwell, G.J., LopezHiguera, J.M. and Tatamm, R.P. (2005) Modification of the Refractive Index Response of Long Period Gratings Using Thin Film Overlays. Sensors and Actuators B: Chemical, 107, 738-741.
30. Wang, Z., Heflin, J.R., Stolen, R.H. and Ramachandran, S. (2005) Analysis of Optical Response of Long Period Fiber Gratings to Nm-Thick Thin-Film Coatings. Optics Express, 13, 2808-2813.
31. Villar, I.D., Matías, I.R. and Arregui, F.J. (2005) Optimization of Sensitivity in Long Period Fiber Gratings with Overlay Deposition. Optics Express, 13, 27-69.
32. Cusano, A., Iadicicco, A., Pilla, P., Contessa, L., Campopiano, S. and Cutolo, A. (2006) Mode Transition in High Refractive Index Coated Long Period Gratings. Optics Express, 14, 19-34.
33. Yang, J., Yang, L., Xu, C.Q. and Li, Y. (2007) Optimization of Cladding-Structure-Modified Long-Period-Grating Refractive-Index Sensors. Journal of Lightwave Technology, 25, 372-380.
34. Korposh, S., Selyanchyn, R., Yasukochi, W., Lee, S.W., James, S.W. and Tatam, R.P. (2012) Optical Fiber Long Period Grating with a Nanoporous Coating Formed from Silica Nanoparticles for Ammonia Sensing in Water. Materials Chemistry and Physics, 133, 784-792.
35. Villar, I.D. (2015) Ultrahigh-Sensitivity Sensors Based on Thin-Film Coated Long Period Gratings with Reduced Diameter, in Transition Mode and near the Dispersion Turning Point. Optics Express, 23, 8389-8398.
36. Sim&otildees, E., Abe, I., Oliveira, J., Fraz&atildeo, O., Caldas, P. and Pinto, J.L. (2011) Characterization of Optical Fiber Long Period Grating Refractometer with Nanocoating. Sensors and Actuators B: Chemical, 153, 335-339.
37. James, S.W., Korposh, S., Lee, S.W. and Tatam, R.P. (2014) A Long Period Grating-Based Chemical Sensor Insensitive to the Influence of Interfering Parameters. Optics Express, 22, 8012-8023.
38. Tripathi, S.M., Bock, W.J., Mikulic, P., Chinnappan, R., Ng, A., Tolba, M. and Zourob, M. (2012) Long Period Grating Based Biosensor for the Detection of Escherichia coli Bacteria. Biosensors and Bioelectronics, 35, 308-312.
39. Zou, F., Liu, Y., Deng, C., Dong, Y., Zhu, S. and Wang, T. (2015) Refractive Index Sensitivity of Nano-Film Coated Long-Period Fiber Gratings. Optics Express, 23, 1114-1124.

40. Yang, R.Z., Dong, W.F., Meng, X., Zhang, X.L., Sun, Y.L., Hao, Y.W., Guo, X.C., Zhang, W.Y., Yu, Y.S., Song, J.F., Qi, Z.M. and Sun, H.B. (2012) Nanoporous TiO2/Polyion Thin-Film-Coated Long-Period Grating Sensors for the Direct Measurement of Low-Molecular-Weight Analytes. Langmuir, 28, 8814-8821.
41. Korposh, S., James, S., Tatam, R. and Lee, S.W. (2013) Chapter 4: Optical Fiber Long-Period Gratings Functionalised with Nano-Assembled Thin Films: Approaches to Chemical Sensing. In: Cuadrado-Laborde, C., Ed., Current Trends in Short- and Long-Period Fiber Gratings, InTech, 63-86.
42. Wei, X.T., Wei, T., Xiao, H. and Lin, Y.S. (2011) Terbium Doped Strontium Cerate Enabled Long Period Fiber Gratings for High Temperature Sensing of Hydrogen. Sensors and Actuators B: Chemical, 152, 214-219.
43. Chen, H. and Gu, Z. (2014) Filtering Characteristics of Film-Coated Long Period Gratings Operating at the Phase Matching Turning Point. Optik, 125, 6003-6009.

CHAPTER 5

An Electron Beam Profile Instrument Based on FBGs

Dan Sporea [1,*], Andrei Stăncălie [1], Nicu Becherescu [2], Martin Becker [3] and Manfred Rothhardt [3]

[1] National Institute Laser, Plasma and Radiation Physics, 409 Atomiştilor St., Măgurele, RO-077125, Romania

[2] Apel Laser, 15 Vintila Mihăilescu St., Bucharest, RO-060394, Romania

[3] Leibniz Institute of Photonic Technology, Albert Einstein Str. 9, Jena 07745, Germany

ABSTRACT

Along with the dose rate and the total irradiation dose measurements, the knowledge of the beam localization and the beam profile/energy distribution in the beam are parameters of interest for charged particle accelerator installations when they are used in scientific investigations, industrial applications or medical treatments. The transverse profile of the beam, its position, its centroid location, and its focus or flatness depend on the instrument operating conditions or on the beam exit setup. Proof-of-concept of a new type of charged particle beam diagnostics based on fiber Bragg gratings (FBGs) was demonstrated. Its operating principle relies on the measurement of the peak wavelength changes for an array of FBG sensors as function of the temperature following the exposure to an electron beam. Periodically, the sensor irradiation is stopped and the FBG are force cooled to a reference temperature with which the temperature influencing each sensor during beam exposure is compared. Commercially available FBGs, and FBGs written in radiation resistant optical fibers, were tested under electron beam irradiation in order to study their possible use in this application.

KEYWORDS

fiber Bragg grating; electron beam; on-line beam profilometry; radiation-hardened optical fiber

1. INTRODUCTION

Along with the dose rate and the total irradiation dose measurements, the knowledge of the beam localization and the beam profile/energy distribution in the beam are parameters of interest for charged particle accelerator installations when they are used in scientific investigations, industrial applications or medical treatments. The transverse profile of the beam, its position, its centroid location and its focus or flatness depend on the instrument operating conditions or on the beam exit setup.

Various methods have been proposed to evaluate the beam transverse profile: arrays of Faraday cups [1,2]; ionization chamber and imaging plates [3]; micro strip metal detector [4]; pepper-pot device, slit-grid method [5] or rotating slits [6]; Optical Transition Radiation (OTR), Secondary Emission Monitors (SEM) [7,8]; phosphor afterglow [9]; scintillating screen [8,10] or scintillating gas detector [11] and charge coupled device (CCD) camera; flat panel detectors [12], arrays of p-i-n diodes [13] or IonCCD [14]; moving wire scanners [15,16] or vibrating wire scanners (VWS) [17]; Compton-scattered laser light [18]; mesh of scintillating optical fibers [19]; detection of Bremsstrahlung radiation scattered by special foils [20].

The present paper reports the proof-of-concept (the design and the implementation) of a beam profile instrument for electron beams, based on an array of fiber Bragg gratings (FBGs). The behavior of commercially available FBGs and FBGs developed in radiation-hardened optical fibers was investigated for the first time under electron beam irradiation, for possible inclusion in the instrument for electron beam profilometry. Tests were carried out under electron beam irradiation at the National Institute for Laser, Plasma and Radiation Physics.

2. OPERATING PRINCIPLE

In the case of FBGs one or several gratings are inscribed in the optical fiber core by various means able to modulate the fiber-refracting index. As it has a broad spectral band, an optical signal propagates along the optical fiber, a narrow, specific wavelength is reflected, and all other wavelengths are transmitted. This way, the grating acts as a very selective wavelength filter. If a FBG is subjected to external factors (temperature, strain), the period of the grating changes, which implies a modification of the reflected wavelength. The shift of this peak wavelength can be linked to the magnitude of the external stimulus [21]. Under these conditions, the grating operates as an intrinsic optical fiber sensor. The two mentioned effects induce a change in the peak wavelength simultaneously, so, in order to monitor only one of the external parameters, the two effects have to be decoupled.

Exposed to radiation, optical fiber gratings undergo a significant or less significant change in their parameters, depending on the type of the optical fiber (photosensitive optical fibers with various dopants or hydrogen-loaded [22]), the technology used to produce the grating (Type I, Type II, Type IIA FBGs [22], long-period gratings [23,24]) or the characteristics of the irradiation field (gamma-ray [22], neutron [25], proton beams [26]). When such gratings are irradiated under high total irradiation doses they show remarkable radiation resistance [27] or demonstrate some vulnerability to ionizing radiation [28], recommending them for radiation dosimetry.

Based on FBGs' response to temperature under irradiation we designed an instrument for the evaluation of electron beam transverse profile in real time. The paper describes the construction and operation of the developed beam profile instrument. According to our knowledge, this is the first report on FBGs exposed to electron beam irradiation.

3. IMPLEMENTATION AND RESULTS

The basic idea of the instrument relies on the wavelength change of FBGs exposed to an electron beam as their temperature increases because of the energy they receive from the incident charged particles. Statistically, spatial mapping of the temperature change is associated with the spatial distribution of the beam energy. In this way, two-dimensional (2D) temperature mapping can be used to evaluate the transversal characteristics of the beam (shape, uniformity, flatness, localization, hot spots detection, etc.). Figure 1 presents a schematic drawing and the practical implementation of the equipment as it was designed to operate, for vertical irradiation geometry in conjunction with the linear accelerator (LINAC) machine developed at the National Institute for Laser, Plasma and Radiation Physics.

From a mechanical point of view, the profile meter has two distinct horizontal planes: in the upper one, a mechanical shutter can be moved along the longitudinal axis of the structure, while the lower plane constitutes the detection plane. In Figure 1c,d the shutter is in the open position enabling the electron beam, which is perpendicular to the two planes from above, to reach the detection plane where the FBGs sensor array is located. The shutter includes a 20 mm thick Al board designed to stop the electron beam from traveling towards the detection plane when closed.

The 2D temperature sensor is located in the detection plane. It includes an array of FBGs embedded into a polystyrene backplane in order to thermally isolate the individual sensors from one another and to prevent any strain from being applied to the gratings during irradiation. The sensors are placed into an array format and are read sequentially by a four-channel sm125 Micron Optics interrogator. The sensor array is exposed periodically to the electron beam, until its temperature reaches a limiting value established by the operator. The upper limit of the temperature depends on sensor characteristics (maximum operating temperature, the dynamic range of the wavelength change with temperature, etc.). The limiting temperature value is detected either by monitoring the maximum wavelength shift of all the gratings or with a thermocouple which is simultaneously exposed to the charged particle beam. The mechanical shutter blocks the electron beam incident on the sensor array as the limiting temperature is reached. This way, no more energy is transferred towards the gratings, which are subjected to an accelerated cooling process, as a fan blows air over them. The operator has the possibility to set the lower limit of the sensor temperature, either as a wavelength change or as indicated by the thermocouple. After cooling is achieved, the shutter opens and the exposure of the sensing array to the electron beam energy begins.

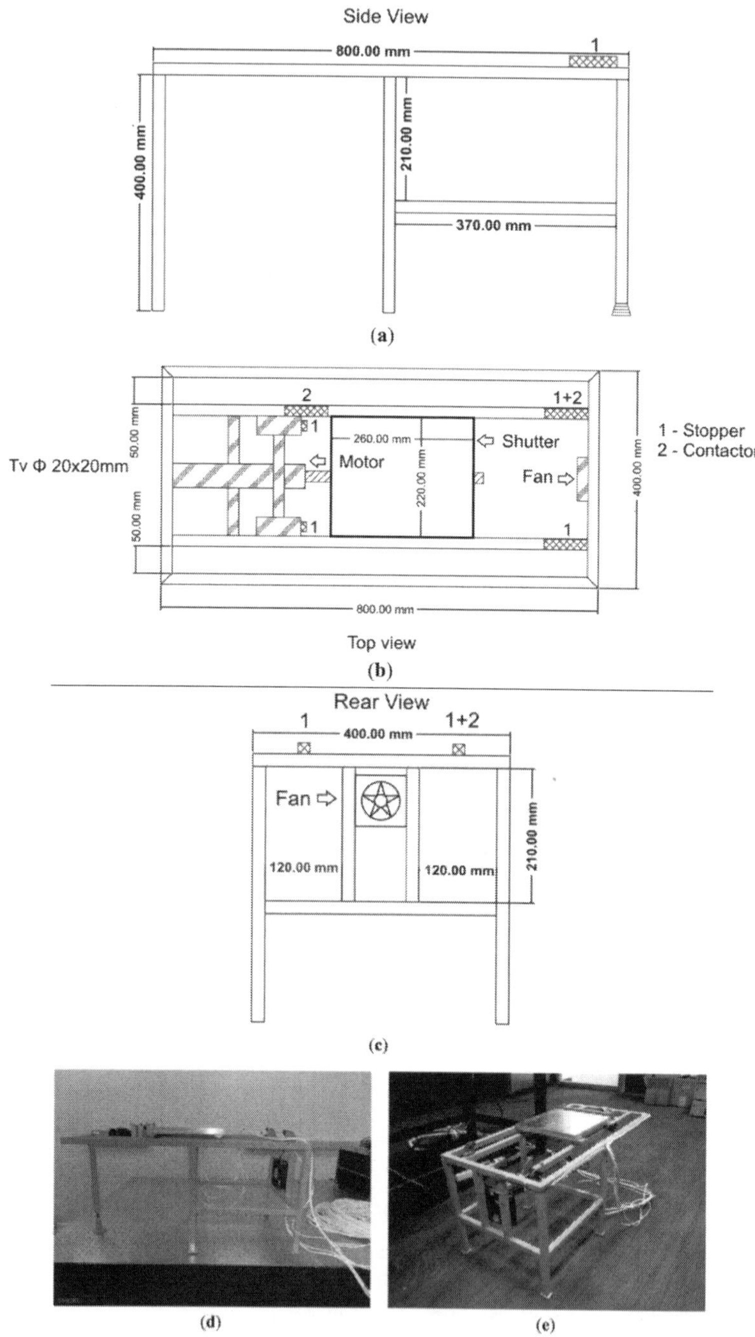

Figure 1. (a–c) The schematic drawing of the electron beam profile instrument based on FBGs: 1—mechanical stops; 2—limiting switches. **(d, e)** the practical implementation of the instrument.

The instrument is controlled by a laptop computer through a specially designed interface and a VI developed in LabVIEW. The limits of how far the shutter opens or closes are detected by the controller based on the signals received from two limiting switches. The extreme positions of the shutter are signaled to the operator on the graphical user interface by two virtual light-emitting diodes (LEDs). The cycle can run continuously or can be controlled manually by the operator. During the period the shutter is off (open) the sensors are scanned continuously and data regarding their wavelength values are saved on the laptop hard drive (HD). The acquired data are processed to build three-dimensional (3D) arrays containing information on the sensors' position in the detecting plane and the instant value of their peak wavelength. This way, between subsequent "shutter off" and "shutter on" commands, the spatial change of the beam energy in the transverse plane is recorded. At the end of the data acquisition cycle, the controller displays the distribution of the beam energy, as detected by the gratings array. This map indicates the beam energy at every sensor location by measuring at each point the excursion of the peak wavelength between the lower limit reached by the end of the cooling cycle and the final value read at each location at the end of the exposure process.

The "exposure time" is set by the operator as a function of the beam energy, sensitivity of FBGs and the speed of the heating process (dose rate). The Micron Optics interrogator is controlled through an Ethernet connection, while the temperature from the thermocouple is collected with a NI-cRIO-9211 temperature data logger by a USB connection. The acquisition rate in our case is conditioned by the interrogator scanning frequency (1 Hz), the number of channels (four channels in our application) and the time required for processing information associated with each channel (i.e., the number of gratings per channel). Theoretically, the unit can accommodate up to 80 sensors. A larger number of sensors can be interrogated using an external multiplexing unit (i.e., Micron Optics sm041-408 or sm041-416).

The instrument has no moving parts in the detection plane and, if needed, can operate in vacuum, as the sensors connected in a daisy-chain setup can be coupled with the interrogator through an optical fiber, multi-port vacuum feedthrough device [29]. In such a case, the air cooling has to be replaced by thermoelectric cooling. The instrument can be controlled remotely (in our setup the distance is about 15 m apart from the irradiation zone), so that no other components or devices are exposed to irradiation or to high electromagnetic interferences. Only the FBGs, the thermocouple and the mechanical shutter are placed in the irradiation area.

The spatial resolution, which can be achieved with such a profile meter, is set along one axis by the distance between two sensor rows (located 3–5 mm apart) and along the other axis by the length of individual FBGs (which in our case was either 12 mm for commercial the products we used, or 4 mm for the custom designed components). The dynamic range for the beam energy detection with FBGs depends on the span between the low and high temperature values. In our study, we set the low temperature value to 18 °C, while the highest temperature was 65 °C, as allowed by the upper limit imposed by the FBGs used. An extended range can be achieved between 18 °C and 200 °C if high operating temperature gratings are used instead. This dynamic range ensures a linear response of the sensor over the temperature range. The resolution of the acquired signal is 0.1 °C, corresponding to the best spectral resolution provided by the interrogator. The temporal resolution of the equipment is dependent on the electron beam dose rate. In the experimental setup we used (an increase of the FBGs temperature up to 37 °C and the air flow provided by a PC fan) the cooling time was about 1–2 min. The presence of "hot spots" or non-uniformities of the beam energy distribution are locally responsible for the upper temperature limit. The mechanical shutter "turn-on" and "turn-off" times are in the order of 30 s and can be adjusted by the operator according to the experiment needs.

The reason for the inclusion in the setup of the shutter and the fan is to eliminate the bias present after each exposure, as the sensor array needs quite a long time to cool down based on only convection. Figure 2 illustrates this bias when the sensor array is exposed to increased total doses and the cooling is provided only by air convection. In the illustrating curves, the positive slopes correspond to the FBGs exposure to the electron beam, while the negative slopes reflect the FBGs cooling, by natural convection as the irradiation is stopped. As can be noticed from the acquired data, it is difficult to naturally cool the sensor array after the exposure stops. The presence of this biasing level in the wavelength readout leads to a long time interval between exposures, as requested by the natural convection. By closing the shutter, the irradiation of the sensors is stopped and they are simultaneously cooled to the reference temperature of 18 °C. This approach also has the advantage of checking the beam quality by obstructing the beam for very short durations. The force cooling is also required as the temperature in the operating room varies when the accelerator is subjected to various operating conditions. The shutter can be used in the "on" (closed) position as support for the samples to be irradiated, so the operator can switch between sample irradiation and beam quality check.

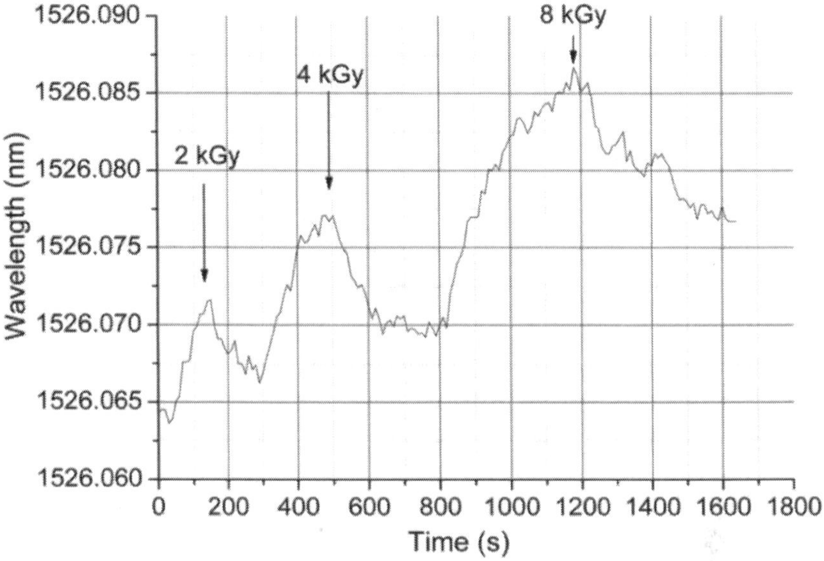

Figure 2. The effect of the bias during the wavelength reading, after subsequent exposure of the same sensor to different doses, in the case of natural cooling by convection.

For all measurements, we considered only the wavelength shift of each sensor against the initial level when the measurements started. The presence of the bias signal along with the imposed upper temperature during the irradiation limits the dynamic range of the instrument. A higher dynamic range can be achieved by: (i) blocking the electron beam simultaneously by force cooling the sensors; (ii) using FBGs able to operate under higher temperatures.

Whatever the situation is, during the exposure of the sensor array, data are acquired continuously at the rate of one scan per second. With this approach, the full "history" of the electron beam energy distribution changes during the monitoring process is available to the operator, who can correlate these changes with the operating conditions initially set (beam current, magnetron voltage, voltage on the injection section of the machine, etc.).

In order to use FBGs as sensors in the proposed electron beam profile instrument and to assess their degradation under this type of radiation exposure we tested some commercially available FBGs and a FBG from a set written in a radiation-hardened optical fiber. Gratings in single mode

(SM) andnon-photosensitive radiation-hardened fibers were written by a two-beam interferometer and deep ultraviolet femtosecond laser radiation [30]. The produced FBGs have a peak wavelength between 1545 nm and 1554 nm and a reflectivity varying between 30% and 90%. The commercially available products are from Technica SA, written in SMF-28 optical fiber, and operating at λ = 1516 nm, 1526 nm, 1536 nm, 1546 nm, each grating having a length of 12 mm, and the following parameters: reflectivity > 80%; BW@3 dB < 0.3 nm; Side Lobe Suppress Ratio (SLSR) > 15 dB; Acrylate recoating.

For the evaluation of the FBGs degradation under electron beam irradiation, tests were done by subjecting the gratings to temperature cycling, heating and cooling them from room temperature to 65 °C, in a temperature-controlled chamber type Memmert UE 400, in order to evaluate their behavior after irradiation. These tests were run initially and between the irradiation steps. During the temperature cycling tests, we monitored the Bragg wavelength change and the full spectrum of the gratings. The chamber temperature was increased linearly via commands from the controller we developed, while the cooling was done through natural convection. For both the heating and the cooling process, the controller acquired data from the NI temperature data logger and from the Micron Optics interrogator. An example of a calibration curve used for the evaluation of the temperature sensitivity for one of the tested FBG is given in Figure 3. After each test, the modifications of the full width at half maximum (FWHM), the temperature sensitivity and the SLSR were estimated. For the total irradiation doses reached during the irradiation campaigns no significant changes of these parameters were noticed.

Figure 4 compares the spectral characteristics of a commercial available FBG before any irradiation and after it was subjected to a cumulated total dose of 292 kGy, both measurements being done under strict temperature control at 23 °C. As can be noticed, even for extensive exposure to the electron beam for over 130 min, no visible degradation of the FBG parameters was observed. The total dose present in the operation state of the instrument we designed was much lower than the dose received during the reliability tests.

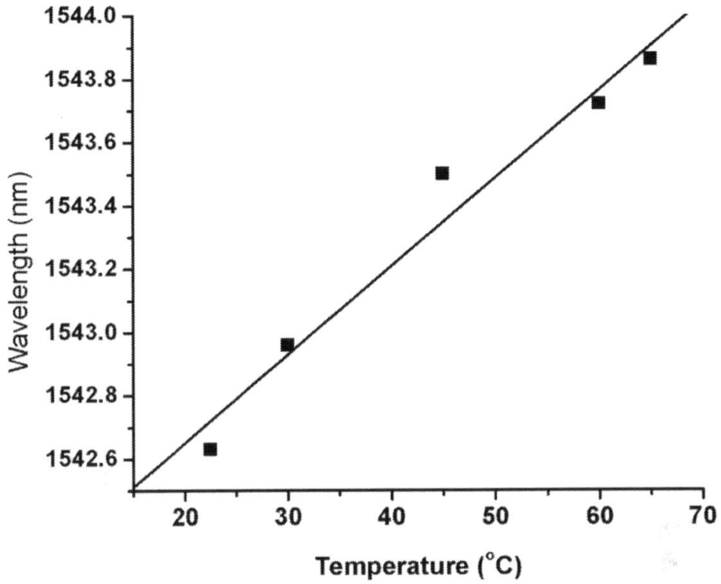

Figure 3. Calibration curve of the fiber Bragg gratings (FBG) sensor.

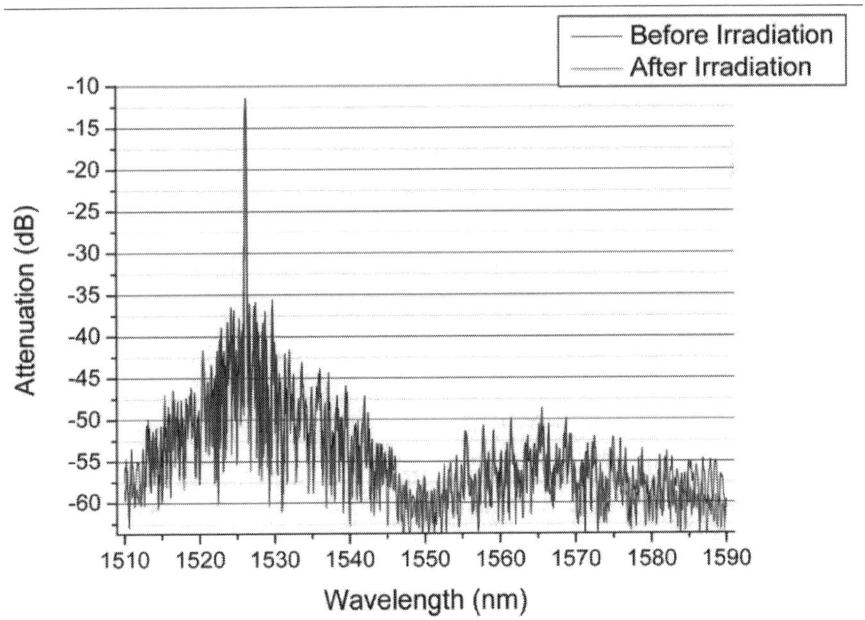

Figure 4. The evaluation of the commercially available FGB parameters: (black) before the irradiation; (red) after exposure to electron beam, measurements done at 23 °C.

The description of the beam profile instrument operating principle and the requirements of the tests performed between the irradiation steps underline the two sets of functions which have to be performed by the controller: (i) to control and to measure the chamber temperature and to acquire the spectrum of a single FBG during the heating-cooling processes; (ii) to collect information on the Bragg wavelength shift from the sensors included in the detection array; to monitor the temperature in the detection plane during the exposure and the cooling of the sensor array; to switch the shutter and the cooling fan "on" and "off" as appropriate; to save the acquired data on HD; to display the recorded data as a 2D/3D representations; to make possible the recall of the stored data.

Based on these needs, the controller assists the operator with both running the instrument as an electron beam profile meter and with testing the possible irradiation-induced degradation of the gratings. To perform these tasks, the controller includes two main units: (i) a custom developed board, a high-side switch 2n06lh5 with the gate driver, controlled by 5 V; a dual full bridge driver able to provide 4 A current to operate a DC motor driving the shutter; the freewheeling diodes for motor overload protection; the 5 V regulator used to monitor the proximity switches states, switches which detect the limits of the shutter travel; (ii) an Arduino UNO board (ATmega328 microcontroller; 32 KB flash memory and 1 KB EEPROM).

The Arduido board supervises the operation of the controller and of the motor, which drives the shutter (sense, speed, detection of the limits). The control of the motor by the Arduino UNO board is done using the LIFA program (LabVIEW Interface for Arduino).

Four graphical windows are available to the operator: "Shutter"; "Thermocouple"; "Oven"; "Interrogator Data Acquisition". Seven buttons on the main window enable the operator to enter commands: (a) "Start cool down"—overwrites any previous command and starts the air flow over the sensors matrix; (b) "Stop cool down"—overwrites any previous command, moves the shutter into the off position and stops the cooler; (c) "STOP Shutter"—stops the shutter at the current location; (d) "Start/Reset Motor Control & Temperature Measurement"—initiates all instrument functions (temperature recording over time, the cooling-exposure cycle, recording of detected peak values, etc.), and does not affect final displayed data; (e) "Start Acquisition"—initiates the display of the acquired data, in graphical form; (f) "Stop Acquisition"—stops data acquisition from the interrogator;

(g) "STOP"—stops the program running and save all data under a name selected by the operator.

At the end of the exposure period the beam profile instrument displays on the screen the final results corresponding to the energy (associated with the total dose) received at each sensing location during the data acquisition process. These data represent only the change of the temperature recorded by each FBG sensor (as a wavelength shift), measured between the low temperature limit and the final temperature. At this moment, the operator has the choice either to allow the software to continue running the irradiation-cooling cycles or to interrupt the process and display additional frames from those acquired during the irradiation. As the frames are time series coded, the operator has the possibility to visualize the evolution of the beam energy transverse distribution at different moments in time.

Figure 5 illustrates the reconstructed results during the tests, in the case of commercially available FBGs arranged in four rows at a distance of 4 mm from one another. The temperature was monitored with the thermocouple placed in the same position, and various total doses were used to expose the sensor array. By using this detection geometry, a mapping of a large field can be reconstructed. In our present implementation, up to 120 frames are stored and can be recalled. The number of frames depends on the dose rate and exposure time. Data presented in Figure 5 corresponds to the final wavelength shift values reached after exposure to 2 kGy, 4 kGy and 8 kGy. From the acquired data, a 3D animation of the beam transverse energy profile can be generated, illustrating the spatio-temporal variation of the signal generated by the sensor array, as it is subjected to subsequent total irradiation doses. The supplementary file "Animation_3frames.tif" is an example of the electron beam profile temporal change reconstructed for the three doses. For visualization the file has to be open with Internet Explorer.

To evaluate the resolution of the instrument, we have to mention that for the tested irradiation geometry, the maximum wavelength change achieved (180 pm) is associated with 8 kGy. Considering the wavelength accuracy of the interrogator to be 1 pm, for our setup the best detection resolution is about 44 Gy. As proved by the calibration curve in Figure 3, the instrument response is liner up to 65 °C for the FBGs we used. In the presented example, the temperature increases up to 28 °C, which is in the linear range of operation of the instrument.

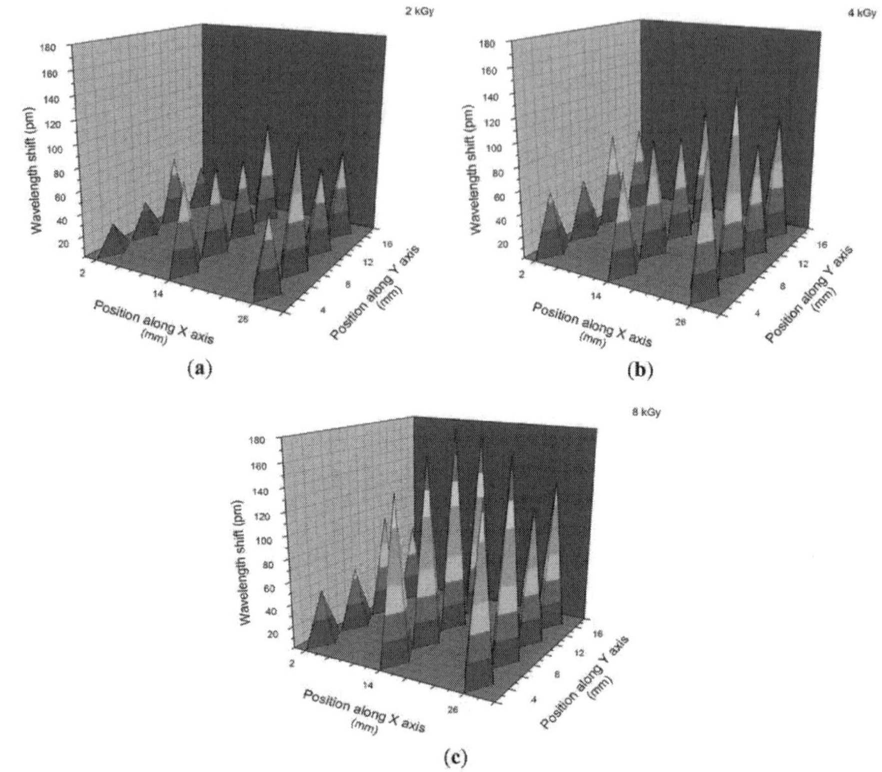

Figure 5. Illustration of the three dimensional (3D) reconstruction of the deployed energy on the FBG array at: (**a**) 2 kGy; (**b**) 4 kGy; (**c**) 8 kGy.

The irradiation tests were run at the Linear Accelerator facility of the National Institute for Lasers, Plasma and Radiation Physics. The electron accelerator, of the traveling wave type, has the following characteristics: pulsed operation; macropulse length—4 μs; mean electron energy—6 MeV; electron beam current—1 μA, pulse repetition rate—100 Hz; pulse duration—3.5 μs, dose rate—2 kGy/min. The beam diameter in the irradiation plane was about 100 mm × 100 mm.

Examples of the results of the irradiation tests carried out on a commercially available FBG and on one FBG inscribed into a radiation-hardened optical fiber are provided in Figure 6. They are used here to derive some conclusions on the radiation resistance of the optical fibers used in the implementation of the electron beam profile instrument. In both cases, the irradiation was done by switching on and off the electron beam.

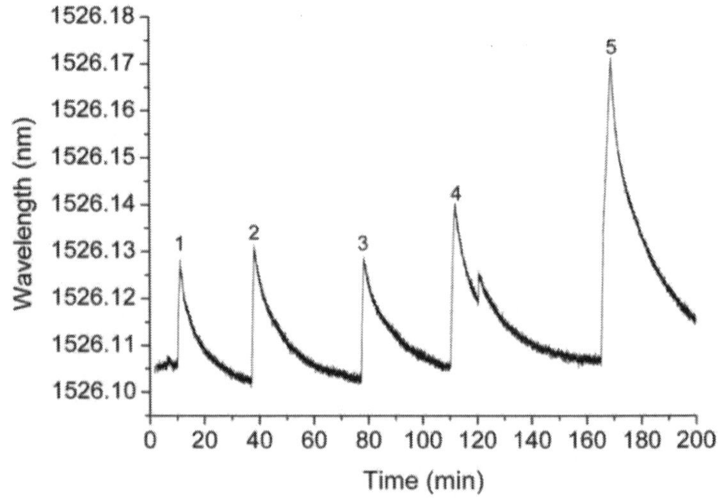

(a)

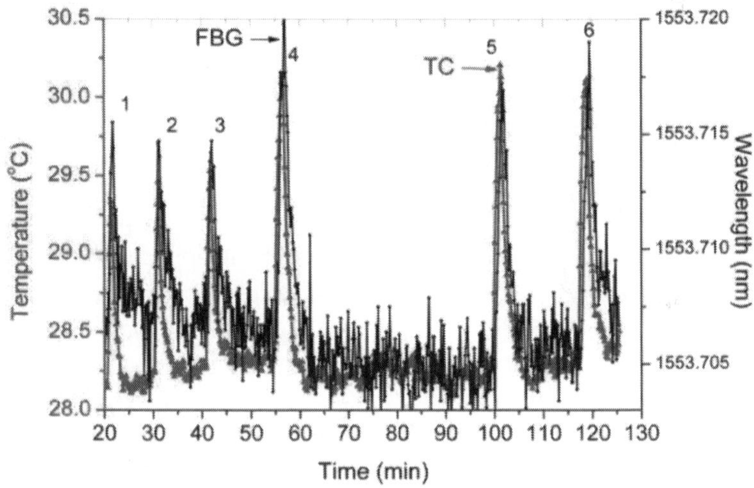

(b)

Figure 6. Results of the exposure to the electron beam of: (**a**) a commercially available FBG; (**b**) a FBG (black curve) inscribed into a radiation-hardened optical fiber (RAD HARD FBG). The temperature in the detection plane was monitored by a thermocouple (TC, red curve).

In the first situation, after the irradiation stops, the cooling of the grating was achieved by natural convection, and no force cooling was employed. For this reason, the wavelength recovery post irradiation is not complete. On the other hand, as the tests were done without any insulation against ambient temperature changes, because the electron beam was propagating in the air, the final temperature reached by the grating was dependent on the ambient temperature variation in the testing facility. This ambient temperature changed as the subsystems composing the LINAC warmed up. In the meantime, the operation of the machine by switching it on and off involved some instabilities related to the transient regime of the accelerator during the start up. For this reason, we adopted a solution with shutter blocking the beam and force cooling. A higher stability and reproducibility of the results can be achieved with this solution. Even operating under natural convection, the detecting system proves a good reproducibility as indicated by the graph in Figure 6a, where points 1, 2 and 3 correspond to a 2.5 kGy total dose, point 4 to 3.5 kGy and point 5 to 6.7 kGy. Nevertheless, according to the proposed implementation, useful information is derived from the difference between the wavelength values at the end and at the beginning of the exposure in every single sensor's location. By their nature, our measurements are relative and not absolute. The proposed method is used for beam profile monitoring, not for radiation dosimetry. This method makes possible the evaluation of the spatial distribution of the beam energy, which can be associated with the total dose involved. The temporal variation of this energy provides information, which can be related to the dose rate.

The thermocouple was used: (i) to monitor the chamber temperature during the calibration process; and (ii) to evaluate the behavior of different types of FBGs during the irradiation tests. Any differences between the FBG and the thermocouple output signals during the exposure to the electron beam signalize a possible degradation of the FGB, as it was irradiated.

The results from Figure 6b indicate a very reproducible recovery of the wavelength inscribed in radiation-hardened optical fibers (RAD HARD) FBG. The wavelength changes strictly follow the temperature variation as indicated by the thermocouple (TC). In this case, points 1, 2, and 3 correspond to a total dose of about 2 kGy, while points 4, 5, and 6 are associated with a total dose of about 4 kGy. The differences between the two cases are explained by the fact that they were not measured during the same irradiation session; however the sensors were placed in the same location. This fact is visible in Figure 5, which clearly indicates the spatio-temporal variation of the incident energy (associated with the dose rate). For the same sensor, the subsequent irradiation and cooling cycles were performed by keeping the sensor in the same position with respect to the incident electron beam. Some differences present in the wavelength shift for approximately the same exposure duration can be explained by the fluctuation of the dose rate of the electron beam. A higher dynamic range is possible if the sensors are cooled down to 15 °C, while the maximum allowed temperature is 65 °C. A higher upper limit of temperature can be associated with longer exposure times, which correspond to higher total doses or dose rates to be monitored.

The problems solved by the proposed equipment are related to the availability of information on the temporal variation of the deposited energy in different locations of the irradiation plane and on the spatial non-uniformity of the energy in the beam cross section. An advantage of our solution is the continuous, multipoint acquisition of data over the beam cross section. By the end of an exposure period, a value proportional to the total energy associated with the total dose received in each sensor location is available (information on beam spatial uniformity is obtained). By recalling intermediary data saved for each location, the operator can assess information associated with the temporal stability of the beam energy.

During the irradiation sessions, the shutter is closed and the samples to be irradiated are placed on the top of the shutter plate. With this setup, when the beam quality has to be checked, the irradiation is stopped for a short time and the beam is sampled by the FBGs array.

4. CONCLUSIONS AND FUTURE WORK

The beam diagnostics method and equipment we present here has several advantages over classical approaches: it is a non-invasive method; it can be operated remotely and under high electromagnetic fields; vacuum operation is easily achieved; daisy-chained and wavelength multiplexing setups make the extension of the number of sampling locations possible; and no moving parts are placed in the detection plane. As our tests have proven, no significant damage was noticed in the FBGs exposed to electron beams. In any case, even if such effects are present, the method accommodates some corrections as only relative temperature changes or wavelength variations, respectively, are recorded for data processing. Besides that, periodical recalibrations are done using the temperature tests run in the temperature-controlled chamber. Being based on an intrinsic optical fiber sensor, the instrument requires no alignment. As compared to some classical methods, the newly proposed one is a real time measurement device. It has the advantage that it almost continuously records the energy distribution in the transversal cross section of the charged particle beam. The intermediary energy distributions can be recalled and compared and associated with the changed operating conditions of the accelerator. The setup by its principle is fire- and explosion-safe to operate. The demonstrated production of FBGs in radiation-hardened optical fibers extends the possibility to operate this beam profile instrument under ionizing radiation conditions.

As these investigations and the associated implementation have acted as a proof-of-concept, we will focus our efforts in the near future to refine the equipment design in order to achieve a better spatial resolution (shorter and closely located FGBs inscribed in the same optical fiber) and to further investigate other types of FBGs, which can be embedded into the detecting system. The use of optical fibers designed for high temperature operation can assure an extended lifetime and a higher dynamic range. In addition, more tests on custom designed FBGs inscribed in a standard optical fiber and radiation-hardened optical fibers have to be performed.

ACKNOWLEDGMENTS

The Romanian authors acknowledge the financial support of the Romanian Executive Agency for Higher Education, Research, Development and Innovation Funding (UEFISCDI), grant 8/2012, project "Sensor Systems for Secure Operation of Critical Installations". The work was done as a cooperation in the frame of COST Action TD1001 "Novel and Reliable Optical Fibre Sensor Systems for Future Security and Safety Applications (OFSeSa)".

AUTHOR CONTRIBUTIONS

Dan Sporea heads the Laser Metrology and Standardization Laboratory at the National Institute for Laser, Plasma and Radiation Physics. He designed and supervised the reported experiments and the design of the instrument, contributed to result interpretation and prepared the manuscript.

Andrei Stăncălie is a research scientist at the National Institute for Laser, Plasma and Radiation Physics. He performed the irradiation tests, processed data and reviewed the manuscript.

Nicu Becherescu, affiliated with Apel Laser, the SME partner of the Romanian national project, contributed to the design of the instrument and the developed the controlling program.

Martin Becker is an operator for fiber Bragg grating inscription. He worked with the deep ultraviolet femtosecond laser inscription setup to achieve fiber Bragg grating based sensor arrays in non-photosensitive and germanium-free fibers.

Manfred Rothhardt is the group leader for the passive optical fiber modules group. The activities of the group comprise the manufacturing of fiber Bragg gratings and their applications. He is responsible for the availability of the inscription laser setup, which was used for the sensor fabrication.

CONFLICTS OF INTEREST

The authors declare no conflict of interest.

REFERENCES

1. Asano, K.; Hosono, Y.; Kohno, Y.; Kohyama, A. Multi-Faraday-cup-type beam profile monitoring system for a dual-beam irradiation facility. Nucl. Instrum. Meth. Phys. Res. Sect. B Interact. Mater. At. **1986**, 17, 186–188. [

2. Koehn, J.A.; Russell, D.A. Charged Particle Beam Profile Measurement. U.S. Patent, 7875860B2, 25 January 2014.

3. Yasuda, N.; Konishi, T.; Matsumoto, K.; Yamauchi, T.; Asuka, T.; Furusawa, Y.; Sato, Y.; Oda, K.; Tawara, H.; Hieda, K. Dose distribution of carbon ions in air assessed using imaging plates and ionization chamber. Radiat. Meas. **2005**, 40, 384–388.

4. Pugatch, V.; Borysova, M.; Mykhailenko, A.; Fedorovitch, O.; Pylypchenko, Y.; Perevertaylo, V.; Franz, H.; Wittenburg, K.; Schmelling, M.; Bauer, C. Micro-strip metal detector for the beam profile monitoring. Nucl. Instrum. Meth. Phys. Res. Sect. A Accel. Spectrom. Detect. Assoc. Equip. **2007**, 581, 531–534.

5. Forck, P. Lecture notes on beam instrumentation and diagnostics. Available online: https://indico.cern.ch/event/162969/contribution/172/material/slides/22.pdf (accessed on 20 August 2014.

6. Vanderberg, B.H.; Cristofolo, M.P.; Wenzel, K.W. Method and Apparatus for Ion Beam Profiling. U.S. Patent, 7064340B1, 20.

7. Wendt, M.; Hu, M.; Tassotto, G.; Thurman-Keup, R.; Scarpine, V.; Shin, S.; Zagel, J. Beam instrumentation for future high intense hadron accelerators at FermiLab. Proceedings of the 42nd Advanced Beam Dynamics Workshop (ABDW-HB′08), Nashville, TN, USA, 25-29 August 2008; pp. 25–29.

8. Harasimowicz, J.; Welsch, C.P.; Cosention, L.; Pappalardo, A.; Finocchiaro, P. Beam diagnostics for low energy beams. Phys. Rev. ST Accel. Beams. **2012**, 15, 122801:1–122801:16.

9. Zhan, T.Z.; Xu, C.N.; Yamada, H.; Terasawa, Y.; Zhang, L.; Iwase, H.; Kawai, M. Beam profile indicator for swift heavy ions using phosphor afterglow. AIP Adv. **2012**, 2, 032116.

10. Boon, S.N.; Van Luijk, P.; Böhringer, T.; Coray, A.; Lomax, A.; Pedroni, E.; Schaffner, B.; Schippers, J.M. Performance of a fluorescent screen and CCD camera as a two-dimensional dosimetry system for dynamic treatment techniques. Med. Phys. **2000**, 27, 2198–2208.

11. Seravalli, E.; de Boer, M.; Geurink, F.; Huizenga, J.; Kreuger, R.; Schippers, J.M.; van Eijk, C.W.E.; Voss, B. A scintillating gas detector for 2D dose measurements in clinical carbon beams. Phys. Med. Biol. **2008**, 53, 4651–4665.

12. Martišíková, M.; Hesse, B.M.; Nairz, O.; Jäkel, O. Test of an amorphous silicon detector in medical proton beams. Nucl. Instrum. Meth. Phys. Res. Sect. A Accel. Spectrom. Detect. Assoc. Equip. **2011**, 633, S259–S261.

13. Palni, P.; Hoeferkamp, M.; Taylor, A.; Vora, S.; McDuff, H.; Gu, Q.; Seidel, S. A method for real time monitoring of charged particle beam profile and fluence. Nucl. Instrum. Meth. Phys. Res. Sect. A Accel. Spectrom. Detect. Assoc. Equip. **2014**, 735, 213–217.

14. Hadjar, O.; Johnson, G.; Laskin, J.; Kibelka, G.; Shill, S.; Kuhn, K.; Cameron, C.; Kassan, S. IonCCD™ for direct position-sensitive charged-particle detection: From electrons and keV ions to hyperthermal biomolecular ions. J. Am. Soc. Mass Spectrom. **2011**, 22, 612–623.

15. Elmfors, P.; Fassò, A.; Huhtinen, M.; Lindroos, M.; Olsfors, J.; Raich, U. Wire scanners in low energy accelerators. Nucl. Instrum. Meth. Phys. Res. Sect. A Accel. Spectrom. Detect. Assoc. Equip. **1997**, 396, 13–22.

16. Liebert, R.B. Method and Apparatus for Ion Beam Centroid Location. U.S. Patent, 4724324, 9 February 1988.

17. Arutuniana, S.G.; Mailian, M.R.; Wittenburg, K. Vibrating wires for beam diagnostics. Nucl. Instrum. Meth. Phys. Res. Sect. A Accel. Spectrom. Detect. Assoc. Equip. **2007**, 572, 1022–1032.

18. Tenenebaum, P.; Shintake, T. Measurement of small electron-beam spots. Annu. Rev. Nucl. Part. Sci. **1999**, 49, 125–162.

19. Søndergaard, C.S.; Hansen, K.; Madsen, D.V.; Rasmussen, J.; Skipper, B.F.; Kristensen, M.; Baurichter, A.; Nielsen, B.R.; Møller, S.P.; Boudreault, G. Charged particle beam profile detector based on Yb-doped optical fibers. Proceedings of the IPAC2011, San Sebastián, Spain, 4-9 September 2011; pp. 1150–1152.

20. Heighway, E.A.; Hohban, J.; Schriber, S.O. Electron Beam Current, Profile and Position Monitor. U.S. Patent, 4059763, 2 November 1977.

21. Hill, K.O.; Meltz, G. Fiber Bragg grating technology fundamentals and overview. J. Light. Techn. **1997**, 15, 1263–1276.

22. Gusarov, A.; Hoeffgen, S.K. Radiation effects on fiber gratings. IEEE Trans. Nucl. Sci. **2013**, 60, 2037–2053.

23. Rego, G.; Fernandez, A.F.; Gusarov, A.; Brichard, B.; Berghmans, F.; Santos, J.L.; Salgado, H.M. Effect of ionizing radiation on the properties of arc-induced long-period fiber gratings. Appl. Opt. **2005**, 44, 6258–6263.

24. Henschel, H.; Hoeffgen, S.K.; Kuhnhenn, J.; Weinand, U. High radiation sensitivity of chiral long period gratings. IEEE Trans. Nucl. Sci. **2010**, 57, 2915–2922.

25. Fielder, R.S.; Klemer, D.; Stinson-Bagby, K.L. High neutron fluence survivability testing of advanced fiber Bragg grating sensors. Proceedings of SPACE TECHNOLOGY AND APPLICATIONS INTERNAT.FORUM-STAIF, Albuquerque, NM, USA; 2004; pp. 650–657.

26. Taylor, E.W.; Hulick, K.E.; Battiato, J.M.; Sanchez, A.D.; Winter, J.E.; Pirich, A.R. Response of germania doped fiber Bragg gratings in radiation environments. Proc. SPIE **1999**, 3714, 106–113.

27. Gusarov, A. Long-term exposure of fiber Bragg gratings in the BR1 low-flux nuclear reactor. IEEE Trans. Nucl. Sci. **2010**, 57, 2044–2048.

28. Krebberl, K.; Henschel, H.; Weinand, U. Fibre Bragg gratings as high dose radiation sensors? Meas. Sci. Technol. **2006**, 17, 1095–1102.

29. Sporea, D.; Sporea, A.; Vâţă, I.; Rusen, I. Optical fiber feed through resistant to pressure differences. Rom. Pat. Appl. A/00733 **2011**.

30. Becker, M.; Fernandes, L.; Rothhardt, M.; Bruckner, S.; Schuster, K.; Kobelke, J.; Frazao, O.; Bartelt, H.; Marques, P.V.S. Inscription of Fiber Bragg Grating Arrays in Pure Silica Suspended Core Fibers. Photon. Technol. Lett. **2009**, 21, 1453–1455.

CHAPTER 6

The Use of Fiber Bragg Grating Sensors in Biomechanics and Rehabilitation Applications: The State-of-the-Art and Ongoing Research Topics

Ebrahim Al-Fakih [1], Noor Azuan Abu Osman [1,*] and Faisal Rafiq Mahamd Adikan [2]

[1] Center for Applied Biomechanics, Department of Biomedical Engineering, Faculty of Engineering, University of Malaya, 50603 Kuala Lumpur, Malaysia

[2] Photonics Research Group, Department of Electrical Engineering, Faculty of Engineering, University of Malaya, 50603 Kuala Lumpur, Malaysia

ABSTRACT

In recent years, fiber Bragg gratings (FBGs) are becoming increasingly attractive for sensing applications in biomechanics and rehabilitation engineering due to their advantageous properties like small size, light weight, biocompatibility, chemical inertness, multiplexing capability and immunity to electromagnetic interference (EMI). They also offer a high-performance alternative to conventional technologies, either for measuring a variety of physical parameters or for performing high-sensitivity biochemical analysis. FBG-based sensors demonstrated their feasibility for specific sensing applications in aeronautic, automotive, civil engineering structure monitoring and undersea oil exploration; however, their use in the field of biomechanics and rehabilitation applications is very recent and its practicality for full-scale implementation has not yet been fully established. They could be used for detecting strain in bones, pressure mapping in orthopaedic joints, stresses in intervertebral discs, chest wall deformation, pressure distribution in Human Machine Interfaces (HMIs), forces induced by tendons and ligaments, angles between body segments during gait, and many others in dental biomechanics. This article aims to provide a comprehensive overview of all the possible applications of FBG sensing technology in biomechanics and rehabilitation and the status of ongoing researches up-to-date all over the world, demonstrating the FBG advances over other existing technologies.

KEYWORDS

fiber Bragg grating; Bragg sensors; biomechanics; rehabilitation

1. INTRODUCTION

In the last decade, fiber Bragg gratings (FBGs) have shown a great potential for applications in the field of biomechanics and rehabilitation engineering due to their prominent advantages such as their small size, biocompatibility, chemical inertness, immunity to electromagnetic interference (EMI) and multiplexing capability [1,2]. These characteristics make FBGs suitable for human body uses that adapt to the sensor material so that they can be used for in vivo measurement and can be left for long-term monitoring [3]. They also offer high-performance alternative, in comparison to standard technologies like electrical strain gauge (ESG), piezoelectric, resistive or other solid-state sensing, either for measuring physical parameters or for performing high-sensitivity biochemical analysis [4].

Presently, FBGs are widely used for specific applications in aeronautics, the automotive industry, structure monitoring in civil engineering and undersea oil exploration [5–7]; in order to measure various parameters such as strain [8], force [9], pressure [10], displacement [11], temperature [12], humidity [13] and radiation dose [14]. On the other hand, most studies towards synergizing optical FBG sensing technology with biomechanics and rehabilitation applications are very recent [3] and have not been commercialized yet. They have been demonstrated for measurement of a wide variety of parameters; including strain inside and on the surface of intact and plated bones, shrinkage stresses in bone cement during polymerization, pressure mapping in orthopaedic joints, stresses in intervertebral discs, deformation in chest wall to study lung biomechanics, pressure distribution in Human Machine Interfaces (HMIs), forces induced by tendons and ligaments, angles between body segments during gait, and many others in dental biomechanics.

To the best of our knowledge, only two articles have partially discussed the general applications of FBG sensing technology in biomechanics [3,15], however, they both did not fully cover all the previously-reported and the very recent and emerging applications of FBGs in the field of biomechanics and rehabilitation engineering. This article aims to provide a comprehensive overview of the applications of FBG technology in biomechanics and rehabilitation in terms of the very recent progress and the status of ongoing researches up-to-date all over the globe. Furthermore, a comparison between FBGs and other conventional technologies is also presented in order to show that FBG technology demonstrates greater feasibility for most of applications in this field.

2. FBG WORKING PRINCIPLES

The FBG involves a spatially periodic modulation of the refractive index along specific region of a fiber's core written in a short segment of a single mode optical fiber with a cladding diameter of 125 μm [16–18]. If light from a broadband source is coupled into an optical fiber containing the FBG, a narrow spectrum is back-reflected and centred around the so-called Bragg wavelength λ_B (Figure 1(a)) [19–22] which depends on the periodic variation Λ of the FBG and the effective refractive index n_{eff} :

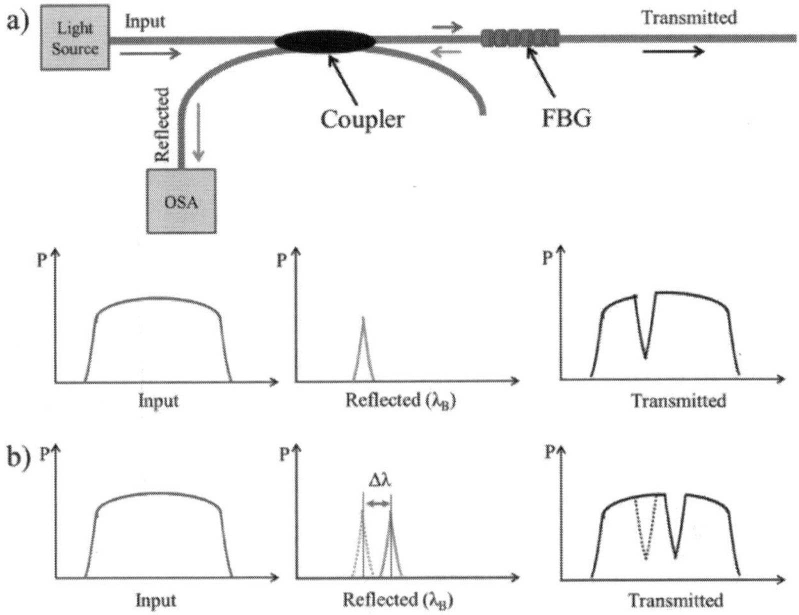

Figure 1. (**a**) The light source (Brown color) is transmitted through the FBG and a narrow band (Green color) is back-reflected, centred around λ_B, and monitored by OSA. (**b**) The back-reflected band is shifted ($\Delta\lambda$) shortly after applying external perturbations.

$$\lambda_B = 2n_{eff}\Lambda \quad (1)$$

When the FBG sensor is subjected to external mechanical or thermal perturbations, the back-reflected peak wavelength will be shifted (Figure 1(b)) according to the extent to which the external perturbations influence the FBG sensor that is spectrally very sensitive when subjected to axial strain [7,23–25]. The axial strain affects the FBG response directly through the compression and expansion changes in the spacing of the periodic variation, Λ, and through the strain-optic effect which induces a change in the effective index of the optical fiber, n_{eff} [2,17,18,20]. The temperature increment ΔT also affects the refractive index and the grating period, resulting in a change in λ_B [22]. The influence of strain and ambient temperature on the grating period and the refractive index can be expressed as follows:

$$\Delta\lambda_B = 2\left[\Lambda\frac{\partial n}{\partial l} + n\frac{\partial \Lambda}{\partial l}\right]\Delta l + 2\left[\Lambda\frac{\partial n}{\partial T} + n\frac{\partial \Lambda}{\partial T}\right]\Delta T \qquad (2)$$

where Δl is the change in FBG length caused by strain and ΔT is the temperature change. To determine the Bragg wavelength λ_B, optical spectrum analyzers (OSA) are commonly used when doing experiments inside laboratories, and, for experiments conducted outside, other specialized interrogators are employed. With high-speed interrogation systems, the FBG sensors would be more suitable for monitoring strains in locations subjected to dynamic loads [20].

3. APPLICATIONS

3.1. Bone

Knowledge of the biomechanical behaviour of the musculoskeletal system is crucial in understanding bone diseases and in designing medical devices [26]. The skeleton normally adapts to mechanical loading, thus understanding of how bone deformation under load occurs is of interest. Several methods have been proposed for the measurement of bone deformation under load. The strain gauge (SG) was the gold standard whose electrical resistance varies proportionally to the amount of strain. A research group based in Israel used SGs for in vivo measurement of strain in human bone in six subjects during running, stationary bicycling, leg presses and stepping, using bone staples made of electrical SGs [27,28]. Recently, several researchers have studied the potential of FBG sensors for in vivo applications as a measurement tool for bone strain, concluding that FBGs have shown competitive advantages over the conventionally-used SGs. FBGs are completely made of biocompatible silicate (SiO_2) glass ceramics materials [26,29] which have a smaller risk for infection, whereas the SG itself represents a foreign body implant. They can also be used in locations where the use of conventional SGs is technically inapplicable [3,26]. For example, when an SG is loaded, its electrical resistance varies proportionally to the applied strain; therefore, it cannot be used in an electromagnetic environment. Moreover, there are other distinguishing advantages of FBGs when compared to the SGs: they can easily adhere to bone and other irregular surfaces [26].

Talaia et al. [30] have determined bone strains in plated and intact synthetic femurs using FBGs and SGs in order to study the effects of fracture fixation plates. For the intact femur, 20 axial SGs were placed on the anterior, posterior, medial and lateral sides at different levels of cortex femur. A simulated 45° fracture on the other femur sample was made and fixed using a 10-hole stainless steel bone plate (Figure 2(a)). Seven FBGs were glued at different locations along the bone plate (Figure 2(b)) and 13 SGs were fixed to the cortex femur (Figure 2(c)). A static load of 600 N was applied to both the plated and intact synthetic femurs and bone strains were recorded at the same sites on the cortex of the femurs and on the bone plate. The sensors have shown good linearity for the different loads. However, FBGs made it much easier to measure strain on bone plates than with the conventional SGs and were more feasible to monitor the evolution of the strain in bone. From the biological point of view, FBGs are able to assess the stiffness of callus formation of fractured bones.

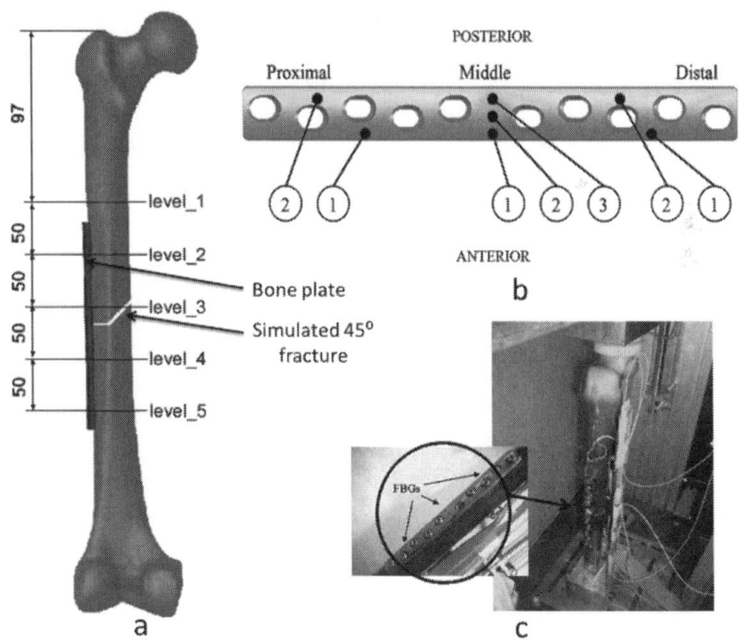

Figure 2. (**a**) The simulated 45° fracture and the locations of the SGs on the intact and plated synthetic Femurs. (**b**) The locations of the FBGs on the bone plate. (**c**) Representation of the experimental apparatus used and detail of bone plate with FBGs [30].

Fresvig et al. [26] have evaluated the use of FBG sensors to detect deformation in human cadaver femur bone under in vitro loading conditions, with potential in vivo application. An acryl tube and a human femoral bone sample of the same length were used in this study. Each sample was instrumented with four polyamide coated FBG sensors and four SGs, interspersed at every 45°, which were placed around the circumference at the central part of each sample. The two samples were loaded using a Mechanical Testing Machine and the induced strains were recorded accordingly. The FBG methodology showed its suitability for dynamic in vitro measurements. Generally, both types of sensors, FBGs and SGs, produced no significant differences, suggesting that the FBG sensors could be successfully implemented for in vivo measurements.

FBG sensors can also be used to understand to what extent the bone calcium loss affects strain response of bone [31]. Two samples of goat tibiae were used; one decalcified in steps while the other remained untreated for a comparative study. Two FBG sensors were directly attached at the midpoint of both samples. Thereafter the two samples were stressed with loads ranging from 0.1 kg to 4 kg using a three-point bending test as shown in Figure 3(a). The strain values were calculated from the FBG wavelength shifts that had been recorded for both samples at the same time in order to understand the effect of decalcification and that of degeneration with time. The strain generated for the untreated sample remained largely unchanged during the experiment, while for the decalcified sample; strain values for same loads increased and became much higher with more calcium loss (Figure 3(b)). By using FBG technique, the strain response gives a direct indication of the degree of calcium present in bone.

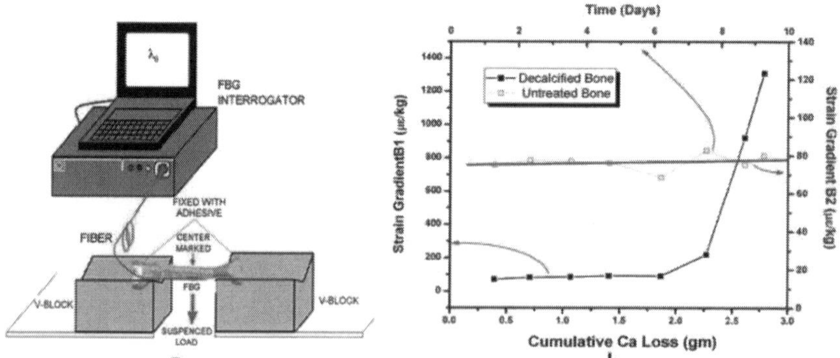

Figure 3. (**a**) The schematic of experimental set-up. (**b**) The comparison between strain response of decalcified and untreated bones [31].

All the above-mentioned studies suggested that FBG sensors have a potential to be used as a novel approach to assess bone biomechanics in vivo. Recently, Carvalho et al. [32] investigated the sensing capability of FBGs when placed in direct contact with human osteoblast cells cultured around the FBG fiber and assessed the reaction of these cells to optical fibers. It was found that osteoblasts exhibited a great capability to adhere and grow over the fiber and the protective polymer coating as well. Furthermore, the FBG sensing capability was not affected throughout the culture period. Hence, these excellent cytocompatibility and sensing capability of optical FBG fibers suggest the possibility of its osseointegration anticipating a variety of in vivo applications in bone mechanical dynamics.

Despite the great ability of bone tissues to adapt to optical fibers, enabling the use of FBG sensors for in vivo applications, there are some obstacles still hampering their use for full-scale implementation in this field. The need of fiber link between the FBG sensing area in the vicinity of patient and the measuring unit is the most common problem [3,15]. In hospitals or mobile care units where this drawback is minimized, portable interrogators could be employed, but for continuous monitoring of patients day-to-day activities, which is an important requirement for orthopaedic treatments and biomechanical studies, these devices can be cumbersome for patients [3,15]. It is possible with today's technology to produce very small wearable interrogators in the near future with no external fiber links, opening up the evolutionary road towards effective in vivo applications of FBG sensors in human body.

In issues pertaining to the influence of total joint replacements on bone strains, a very recent study by Reikeras et al. [33] investigated the differences between internal and external cortical strains in the proximal femur after insertion of a cemented and uncemented hip stem prosthesis using SGs and FBGs. One cadaveric femur was removed within 24 hours from a 64 years old male patient who died from a heart attack. Four SGs were attached at the external femoral cortex both proximally and distally and four optical fibers were sealed at the internal femoral cortex proximally, as shown in Figure 4. When axial load was applied onto the prosthesis, the strains at the proximal level were simultaneously recorded at the internal cortex (by FBGs) and at the external cortex (by SGs) and were then compared in order to evaluate the difference in strain patterns at the internal and external cortices. Previously, it had been assumed that strain on the external cortex reflects that at the internal cortex. However, this study showed that the strain pattern on external cortex was significantly different from that obtained at the internal cortex. In other words, the situation on the external cortex does not correspond to that at the internal cortex and the strain at internal cortex cannot be deducted from strain measured at the external cortex. The ingrowth of bone and stability of the prosthesis depend on the internal cortical surface, therefore, future studies should pay attention to this fact.

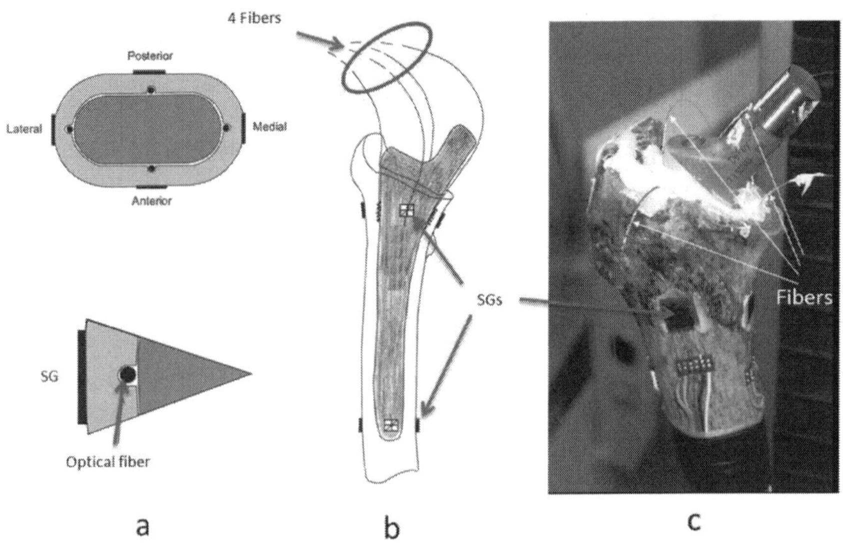

Figure 4. Schematic of the experimental set-up. (**a**) A cross-section of the implanted femur, showing the locations of FBGs and SGs. (**b**) Schematic drawing of the femur with cemented prosthesis, showing the internal (optical fibers) and external (SGs) positions of sensors at the proximal and distal levels. (**c**) Image of the experimental set-up of an implanted cadaveric femur [33].

The strain behaviour at the anterior femoral notching region during total knee replacement and its consequences on the cortex of the distal femur was studied by Completo et al. [34]. They also questioned the effect of using femoral stem in an anterior femoral notching to reduce the fracture risk. It had been hypothesized that femoral notching weakens the cortex of femur which can make it susceptible to fracture in the early postoperative period [35]. In this study, 12 synthetic femurs were selected and undergone experiments under two load scenarios. Femoral components with and without femoral stems were implanted in femurs with different notch sizes to predict experimentally the strain levels at the notch edge using FBGs and at notch region using SGs (Figure 5). The most important finding of this study is that the strain behaviour was dissimilar for the different notch depths. For notch depths lower than 5 mm, the use of stem reduces the strain level at the notch edge to values below the intact femur condition, while for depths greater or equal to 5 mm, the strain levels at the notch edge were higher than the intact femur condition (ranging from +10 to +189%). This suggests the use of a prophylactic stem for notch depths greater than 5 mm, contributing to reducing the fracture risk.

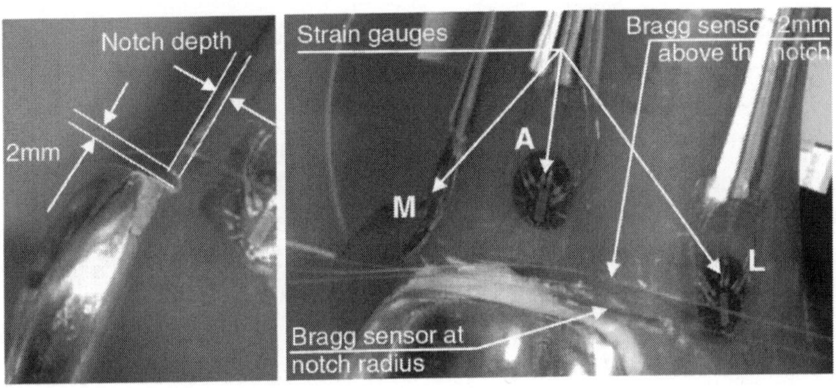

Figure 5. Images depicting the composite femur models with notch. **(left)** The notch dimensions, and **(right)** the positions of SGs (A anterior, M medial, L lateral) and FBGs [34].

3.2. Bone Cement

Damage in bone cement induced by long-term dynamic loading that causes prosthesis loosening is one of the main reasons for total hip replacement failures [36]. In vitro studies using human and animal bones and models could be useful to verify the action of the cement, both during the curing process and after consolidation. Embedded rosette gauges for strain measurement of bone cement in femoral prosthesis had been previously used by Stolk et al. [37]. However, the FBGs have shown their advantages in studies pertaining to these issues. Ramos et al. [38] assessed the accurate stress and strains magnitudes inside cement mantles using FBG sensors which could be useful to predict the mechanical failure. Twelve FBG sensors were bonded to a pre-designed femoral prosthesis; four at the proximal, distal, and half of the stem, and were strategically attached in the anterior, posterior, medial and lateral aspects. The FBG-instrumented prosthesis was placed inside the femur canal in a way the FBGs were placed at a distance of 1 mm from the cement-implant interface. The cement strains measurement using this technique is less time-consuming and easier to implement than the one using SGs.

FBGs can help predict the functional durability of PMMA. Frias et al. [39] tested the capability of FBGs to measure strains inside bone cement during different mechanical tests at real-time. Bone cement was tested at different temperatures and load conditions according with those expected inside the human body. The mechanical behaviour of different specimens of PMMA bone cement was studied in different short-term quasi-static tests to predict the PMMA durability. Different test procedures allowed for measuring the tensile and compression strains on PMMA bone cement using FBG sensors. The embedded FBG was subjected in independent tests to some of the physical parameters that bone cement is simultaneous submit in an artificial joint. The optical reflected spectrum was recorded at the end of each test using an OSA. All the strain measurements inside bone cement reported a linear response that shows a good adaptation and optimization of the load transfer between the PMMA cement and the embedded FBG sensor.

3.3. Orthopaedic Joints

An embedded array of FBGs exhibited its feasibility to be used at orthopaedic joints interfaces for pressure mapping and the measurement of contact forces and stresses. It is necessary for those involved in prostheses research and biomechanics studies to understand the contact area and contact stress at Orthopaedic joints as well as the alignment of the joint interface (i.e., tibiofemoral interface). Measurements of contact area, contact force and stress distribution in the tibiofemoral joint have been made using commercially available stress-sensitive films such as I-Scan and Fujifilm [40–43]. However, the biomechanics of knee joint might be altered after the insertion of these films at the articulating surfaces.

Due to their flexibility and multiplexing capability, FBG sensors have shown an edge over these conventional measurement systems as they are suitable to be used in any irregular-shaped interface surfaces (i.e., knee joint) and, in some cases, they do not change the natural contact topology of joint capsule (i.e., hip joint). The application of an FBG sensor for pressure mapping at the prosthetic knee joints was first reported by Mohanty et al. [44,45]. In these cadaver studies, a rectangular array of FBGs was embedded into a tibial spacer that was inserted at the tibiofemoral interface and the changes in the contact stress distribution caused by joint misalignment in extension and flexion were monitored. This spacer sensor can be used to correct malalignment during total knee arthroplasty. It can be used in vitro for studying the biomechanics of the prosthetic knee joint especially for curved and conforming designs.

The commercial sensitive films were also used for stress measurement in hip joint [46], but their application is severely limited by the amount of dissection required to insert them in this joint. This insertion approach is considered to be too interventional when used in the natural hip [47]. FBG sensors have shown their great potential to overcome some of the sensitive films limitations because they have key advantages that make it more attractive for application in the hip joints such as the small size, mechanical compliance and biocompatibility.

Dennison et al. [48] were the first to report the simultaneous measurement of contact force/stress and fluid pressure in intact cadaveric human hips using Bragg-based sensors, in order to address the limitations shown by the conventionally-used sensitive films. Two 1 mm long uniform FBGs were inscribed in a core of two separate SMF-28 fibers, and were each coated by two layers of polymer annulus and a polyimide sheath, respectively. The polymer layer was placed between the outer diameter of the optical fiber and the inner diameter of the polyimide sheath. Two cadaveric hip samples were obtained and all musculature of the samples was dissected, but the joint capsule was left intact. The FBG contact sensor was placed in the hip joint between the femoral head and the acetabulum cup, while the pressure sensor was put in the fossa which is the active pressure sensing area (Figure 6). Numerical models were used to predict the sensor performance and the predicted sensitivity was verified through experimental calibrations. The results show that contact force and pressure measurements using FBG sensors in the hip joints exhibited repeatability and the polyimide sheath increased the sensitivity of the contact force sensor, when compared to the unsheathed sensor of the same diameter. These sensors could potentially be used to study the correlations between contact forces and pressures in healthy and deteriorated joints.

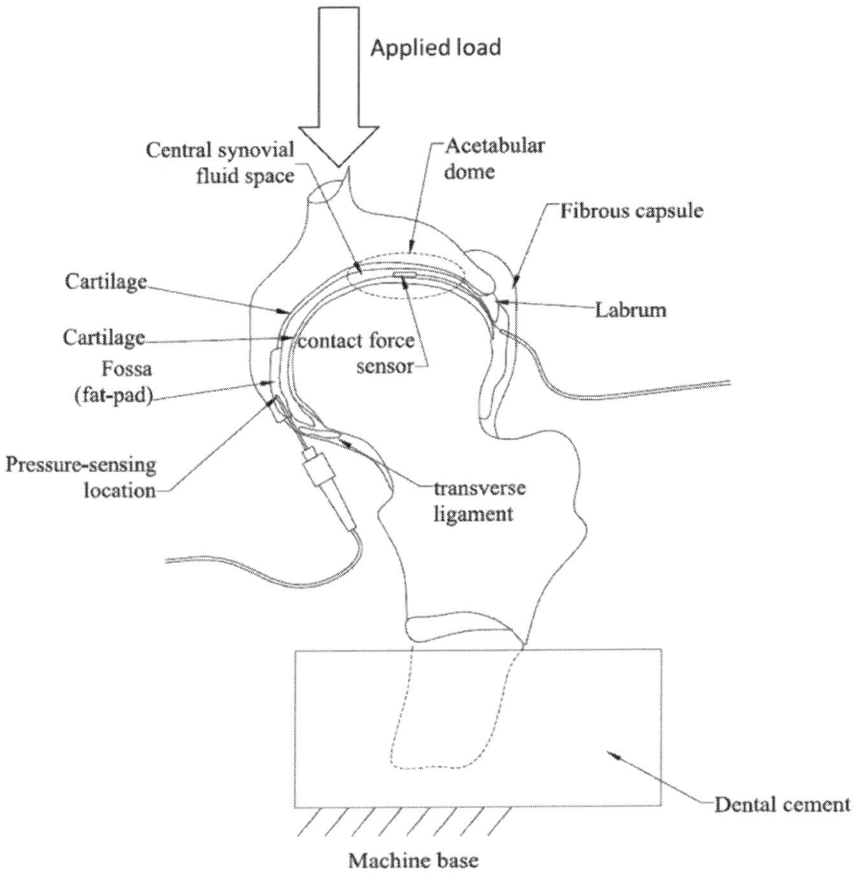

Figure 6. Schematic drawing of the cross-section of hip joint showing the insertion of the pressure and contact stress sensors into their optimal sensing locations. The hip stem was embedded in dental cement for fixation purposes during applying vertical loads [48].

3.4. Intervertebral Disc (IVD)

Pressure distribution at the IVD is a key to understand the IVD biomechanics which were previously measured both in vivo [49,50] and ex vivo [51–54]. Needle-mounted SG sensors have been employed for the measurement of disc pressure [55–59] ranging over 3 MPa in cadaveric human discs [52], but it was found that these sensors show key limitations hampering their use research area. They are housed in large and rigid needles (over 1 mm diameter [60]), which when inserted cause injuries to the disc annular fibres and alter the mechanics of the intervertebral disc [61]. On the other hand, FBG strain sensors have shown excellent ability to address some of the above mentioned limitations as they are smaller in size (less than 0.5 mm) and biocompatible. Due to their small size, they can minimize the injury to the fibers of the disc annulus and the effects on disc mechanics [62]. In other words, using FBGs for pressure measurement is less invasive and disruptive than current measuring systems.

Methods to develop intervertebral pressure sensor using FBGs were first reported by Dennison et al. [63]. A bare FBG sensor (10 mm length and 125 μm diameter) was inserted in a hub of a 25 gauge hypodermic needle which its tip was then passed through the disc annulus so as to be placed approximately at the centre of the nucleus. The sensor was then advanced past the needle tip so that its entire length was exposed to the nucleus pulposus and the needle was left in the annulus to shield the fiber optic cable from any traction. Each lumbar spinal unit was loaded from 0 N to 2,000 N to 0 N at 10 N/s. The sensor performance was investigated in five two-vertebra human cadaveric spinal units. After each test with FBG sensor and for comparison purposes, the tests were repeated by using two SGs instead. Although the pressure measurements using FBG sensors were in agreement with those obtained in previous disc pressure studies, they were not always consistent with measurements made using current standard SGs sensors. That is because bare FBG measurements have shown poor pressure sensitivity, resulting in highly discretized data over the 3 MPa pressure ranges. These drawbacks might come from the nucleus material inhomogeneity and from the length of the FBG (over the 10 mm) which would interrupt pressure measurements. Improving the sensor design and sensitivity and its spatial resolution is very important for obtaining better measurement results.

The same research group have designed and constructed a new patented FBG based IVD pressure sensor [64] that has both improved pressure sensitivity (seven times greater than that of a bare fiber while maintaining extremely small size) and spatial resolution by limiting the sensing region to the tip of the sensing hypodermic tube [1]. A 10 mm length FBG was written in a single mode fiber that was housed within 50 mm stainless steel hypodermic tube so that the FBG is positioned at the end of the tube (Figure 7(a)). A silicone sealant was used to fill up the annular volume between the inner diameter of the hypodermic tube and the outer diameter of the FBG fiber in order to enhance the sensor sensitivity. The sensor tip was introduced into the nucleus center of a cadaveric porcine disc and compressive loads (from 0 N to 500 N at 40 N/s) were applied onto the superior vertebra of the spinal unit (Figure 7(b)). As applied pressure onto the hydrostatic tube is increased from 0 MPa to 3 MPa, the silicone sealant and the optical fiber deflect mainly in the FBG longitudinal axis (compressive strains), with the highest magnitude deflections occurring at the probe tip (the inset of Figure 7(a)). The deflections in the longitudinal axis decrease in magnitude along the sensor (from right to left in Figure 7(a)), and are at their minimum at the left-hand side of the sensor. The same procedures were subsequently repeated using a needle-mounted SG sensor. The measurements obtained using FBG sensor were compared to those made with the widely-used SG sensors. The FBG sensor exhibited higher sensitivity, excellent repeatability and was in agreement to those obtained using SG sensors. It is worth mentioning that this hypodermic tube containing the FBG does not have a cutting tip to ease the insertion into the disc annulus. That is why another cutting tip needle is needed to pierce the annulus prior to insertion. Future design must take this issue into account so as to dispense with the aid of the needle by constructing new FBG sensors with their built-in cutting needle.

The performance of the latter FBG sensor was validated in a subsequent study conducted by Dennison, Wild, Dvorak, Wilson and Cripton [62]. This study assessed the sensor sensitivity, linearity, and hysteresis and determined the sensor accuracy in pressure measurements of IVD as a function of compressive load in six porcine cadaveric spinal units obtained from two lumbar spines. After calibration and preconditioning, the FBG sensor was introduced into the annulus and compressive loads were applied to the functional spinal unit using a saw-tooth loading profile for both loading and unloading phases. The hydrostatic pressure at the disc annulus was monitored using the FBG sensor. Thereafter, the FBG sensor was replaced by SG sensor and same test procedures were conducted. The results of the FBG sensor were still in agreement with those obtained using SG sensors. But the FBG sensor also exhibited higher sensitivity and increased spatial resolution because its active sensing region was confined to the needle tip. What is superior with this FBG sensor over the previous disc pressure sensors is that it did not interfere with vertebral endplates and did not alter the disc biomechanics and structure. In addition, this FBG sensor is 83%

smaller and more flexible than any previously-reported disc pressure sensor and can be used to make less invasive measurements of disc pressure in vivo and during discography.

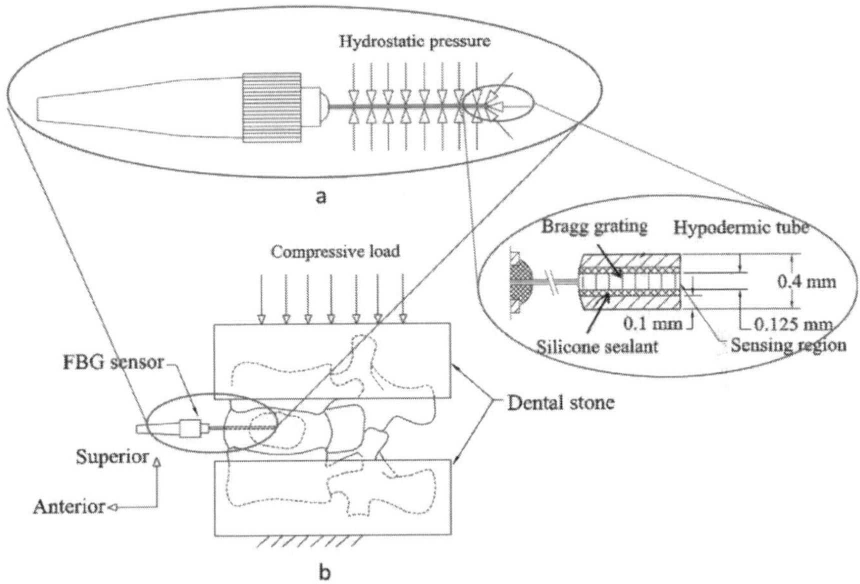

Figure 7. (**a**) Schematic showing the construction of the IVD pressure sensor. The hydrostatic pressure over the hypodermic tube and at the sensing region whose dimensions are shown (right inset). (**b**) Compressive loads applied to a functional spine unit embedded between dental stones and the pressure sensor inserted into the annulus [1].

A very recent study by Roriz et al. [65] has proposed an FBG sensor to measure strain due to bulging of the IVD under axial compression. This study introduced the least invasive sensor up-to-date as the needle was used only for guiding and placing the sensor and could be then removed leaving the sensor in direct contact with the disc material. A 2 mm length FBG was inserted into the disc center of an ex vivo porcine dorsal spinal unit using a 25 gauge hypodermic needle that was removed after placing the sensor and axial compression loads were applied causing strain in the FBG fiber which, combined with strain due to lateral disc bulging, induced changes in the Bragg wavelength. The results verified that the sensor is capable of measuring strain due to the applied loads and the bulging of the annulus fibrosus as a consequence of axial compression. However, this sensor has shown a limitation. Since the FBG fiber is stiffer than the IVD, the fiber will resist disc bulging and fail before maximum disc elongation. Therefore, this technique is not suitable for measurement of tissue physiologic strain.

These new FBG sensors can potentially increase our understanding of the correlations between the internal disc pressure and the degree of disc degeneration. These correlations may increase our knowledge of disc biomechanics and potential disc injuries and degeneration. Further studies are needed to address the effect of the sensor insertion into the disc on the load transfer and the pressure measurements.

3.5. Dental Biomechanics

The application of FBG sensors in the area of dental biomechanics had been investigated earlier. Tjin et al. [66] have reported the first application of FBG sensors to monitor the force and temperature as a function of time in dental splints worn by patients suffering from sleep apnoea. Due to the small size of FBG sensors, it is much easier to embed them within the splint while maintaining the splint efficiency unaltered. These sensors can determine the correct use of splints when worn by the patient. Two pre-calibrated FBGs inscribed in a single strand of an optical fibre were used; one embedded within three layers of glass fibre composite materials to insulate the FBG pressure sensor from temperature effect and the other bare FBG for temperature sensing. The sensors were carefully embedded into the patient's dental splint; the pressure sensor located at the first molar and the temperature sensor located further in, where there is minimum contact with the gums (Figure 8). As the splint is being positioned in the patient's mouth, the results show an increase in both temperature and force. The advantages of these sensors, besides their effectiveness in monitoring the proper usage of the dental splints, are that they are unnoticeable due to their small size and multiplexing capability, and inherently safer than the conventionally used methods (i.e., thermistors and piezo-electric pressure sensor) due to their immunity to EMI.

Another important application of FBGs in dental biomechanics is to measure the properties of dental materials like the stress-strain pattern. A few studies were reported on the possibility to use FBG sensors to study dental materials such as resin-based composites and gypsum products. FBGs can be used to measure the hardening of dental cement and the corresponding stress build-up and the volumetric shrinkage of the resin cement during polymerization [67–69]. Milczewski et al. [70] have used FBGs to determine the polymerization contraction and setting expansion of three types of gypsum products (plaster, dental stone, and high strength dental stone) and to monitor the strain evolution and temperature during the drying phase of the material. Two FBG sensors were used; one sensitive to the strain and temperature variations during setting, and the other placed in a double needle to monitor only the temperature variations during the reaction. The results show that the plaster of Paris demonstrated higher expansion than orthodontic and dental stone. The FBG technique can be a good tool for dentists to better manipulate a material and predict how it will behave in vivo.

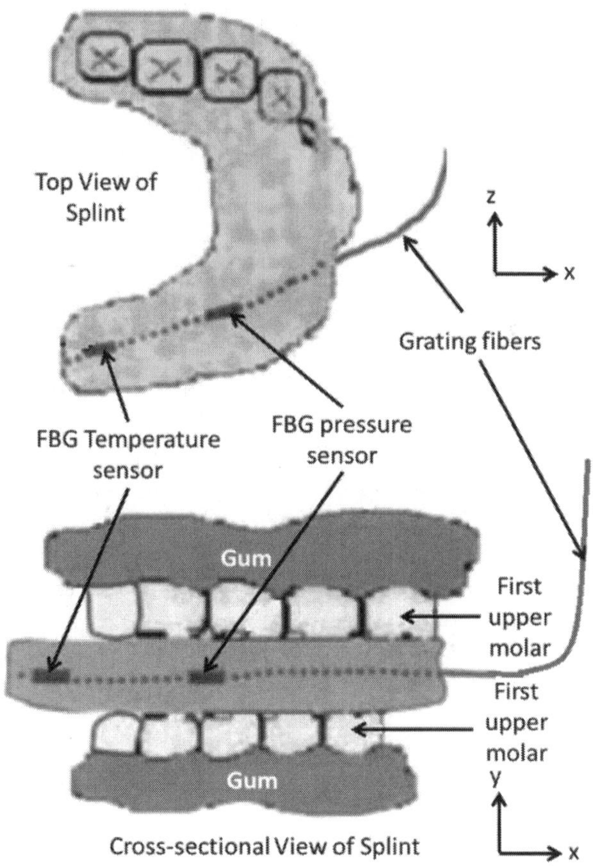

Figure 8. Schematic diagram of dental splint with FBG embedded sensors [66].

A research group in Portugal has employed FBGs to measure strains at the surface of an implanted cadaveric mandible caused by impact loads. Although they had previously used SGs in a few studies pertaining to the mandible [71,72], they studied the potential use of FBG as a biomechanical sensor [73]. In one of their studies [74], they measured the dynamic strains at the mandible surface induced by impact loads. A standardized dental implant was implanted in a dried human cadaveric mandible embedded in a resin support, and an FBG was placed on the outer surface of the mandible near the canine tooth position in a direction parallel to the implant longitudinal axis. The impact loads were generated by striking the implant with a steel mass (52 g) dropped inside a steel tube. The obtained results agreed with those previously reported studies using SGs, showing the ability of FBG sensors to measure dynamic strain variations, with precision, in complex biomechanical systems.

In a subsequent work by the same group [75], they have measured the strains at the mandible surface due to both static and impact loads, using FBGs and SGs, and two standardized implants were implanted in the cadaveric mandible used in the previous study. Two uncoated FBGs and two uniaxial SGs were placed on the mandible surface in the longitudinal axis of the implants (One FBG and SG near each implant), as shown in Figure 9. For the static loads, the implant was step loaded up to a maximum of 160 N, and strains at the surface were recorded by each sensor, FBG and SG. The impact loads were generated in a manner similar to that used in the previous study [74] which was done by striking the implant with 52 g a steel mass. The results of both types of sensors were compared, showing very similar response when both static and dynamic loads were applied. The FBG sensor exhibited better sensitivity with small strains compared to SG, showing its feasibility to be used as an alternative method for measuring strains in dental biomechanical applications.

FBGs can also be used to study model mandibles that replicate the mechanical properties of human bone. A synthetic model was used to study the strain distribution over the mandible branch and articulation caused by the action of the muscles [76]. Four FBGs were glued at the outer surface of the mandible and the strain patterns were assessed for different load configurations including the forces of the masseter and temporal muscles and occlusion loads on different teeth. The same device was modeled by Finite Element (FE) method to determine the strain at the mandible surface. The strains obtained from the FBGs were correlated to identical ones obtained with a numerical FE model. The comparison of the results is considered satisfactory. For symmetric loading the strains measured were identical. For the non-symmetric loading, the FE boundary conditions must be simulated more accordingly. Such results validate numerical methods used in the design of mandible and condoyle prosthesis.

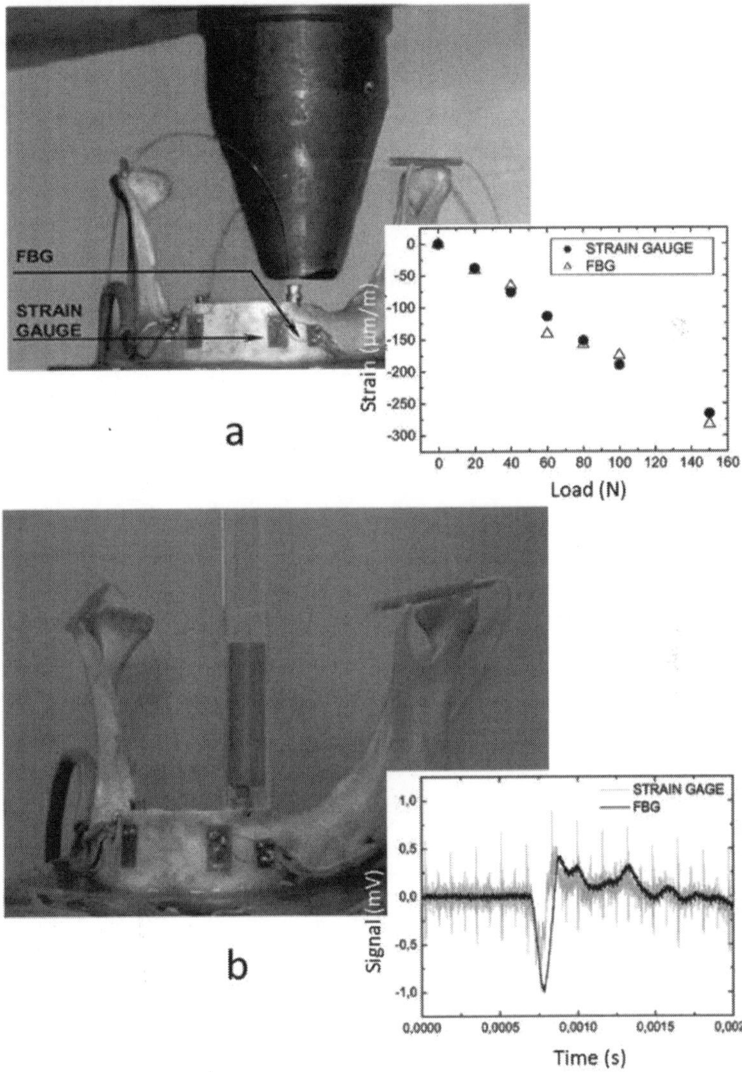

Figure 9. (**a**) The set-up of the step loads. The FBGs and SGs attached to a cadaveric mandible. The inset illustrating the strains measured by both techniques. (**b**) The set-up of the measurement of dynamic strains and the inset showing the FBG versus SGs strains in time-domain [75].

Measuring strains in the mandible bone helps predict the load transfer from the implant to the surrounding bone, leading to design novel dental implants that minimize the strains induced in the mandible. Macro-models of dental implants were studied with FBG sensors inserted in real bone using samples from bovine femur [77,78]. Three types of dental implants; metallic, polymeric and polymer coated metal implants, were compared in order to study the load transfer from the implant to the bone. The FBGs were inserted into the bone in a direction parallel to the implant axis and the static and dynamic strains were recorded as a function of loads. The results show that polymer coated implants can be more appropriate due to better load transfer. FEM analysis can be performed in order to compare and validate the experimental results.

FBG sensors have been used to measure the forces applied onto the surface of teeth and the consequent displacement caused by the orthodontic systems. Low magnitude applied forces caused by orthodontic systems are considered more effective for moving the teeth than impulsive loads. Therefore, Milczeswki et al. [79] investigated the forces applied onto the surface of an incisor by orthodontic system using FBG technology. The experiments were performed in an artificial maxilla with dental wax and metal teeth and the teeth were instrumented with a fixed orthodontic appliance. An FBG imprinted in a HiBi fiber was placed between the incisor surface and the orthodontic bracket, and loads were applied representing the auxiliary appliance. The anterior-posterior force and displacement of the tooth movement were measured, showing that the small size and the high sensitivity of the FBG sensor allow for the determination of forces applied orthogonally to the surface of the teeth. This technique is promising and can be used in the other types of orthodontic applications for in vivo experiments.

Recently, Milczewski et al. [80] went a step further and applied a similar approach using FBGs to study the magnitude and location of forces at the teeth roots and maxilla bone during applying load by orthodontic appliance. A maxilla model was instrumented with FBGs, three pairs of FBGs positioned along three artificial teeth roots, immersed in the elastomeric material, and the other four multiplexed fiber sensors placed transversally along the teeth apex. Preconditioning tests were performed to evaluate the sensitivity of the Bragg gratings in-situ. After the activation of the orthodontic appliance, the teeth were loaded and the strain on the maxillary bone and dentition was monitored showing a linear relationship between the strain and the applied load in the incisor, canine and molar teeth. This also can show the transference of the forces to the teeth and finally to the internal maxillary structures.

In a recent (2011) study by Tiwari et al. [81], FBG sensors were utilized to monitor the impact absorption capability of mouthguards. To assess the mouthguard efficiency when subjected to impact loads, the strain transfer to the teeth through a custom-made mouthguard was studied. Two sets of FBGs were used, one placed at the anterior region and another at the posterior region. Each set has two FBGs; one was attached to the jaw model and the other on the mouthguard surface at parallel locations so that both FBGs are simultaneously affected by the impact (Figure 10(a)). Impact was made using a pendulum device with interchangeable round objects, i.e., cricket ball, hockey ball, and steel ball, and strain induced due to each impact was determined from the FBG wavelength shifts. The wavelength shifts for the FBG bonded on the jaw model exhibited much lower than the shifts for the FBG bonded on the mouthguard (Figure 10(b)), indicating that most of the impact energy is absorbed by the mouthguard. This study can help design better custom-made mouthguards for different malocclusion, as the FBG technique can give strain not only at the point of impact but at other vulnerable points also in a single hit.

3.6. Chest Wall Deformation

The monitoring of chest wall movement during respiration is crucial in understanding lung biomechanics. The chest displacement due to respiration normally ranges from 4 to 12 mm [82,83]. Sensors composed of a bent optic fiber for the measurement of respiratory chest circumference changes had been previously developed by Babchenko et al. [84]. This sensor was composed of a bent optic fiber connected to a chest belt so that its radius of curvature would change during respiration due to changes in the respiratory chest circumference and the light transmission was measured through the bent fibre providing information on the curvature changes. However, this method was replaced with FBG technique which was first reported by Wehrle et al. in 2001 [85]. In this study, the measurement of dynamic strain on human chest during respiration was developed and tested. The FBGs were embedded into a chest deformable strap which was strained in accordance with the deformation of chest. This sensor showed its ability to detect deformation in the thoracic cage during respiration, indicating the respiratory frequency spectrum. It can also be used to trigger a device for inducing electrically assisted ventilation, when one lung is to be excited in phase with the other working lung, or to monitor high frequency oscillatory ventilation. Moreover, the FBG sensors showed their capability to measure the muscle movement in the chest.

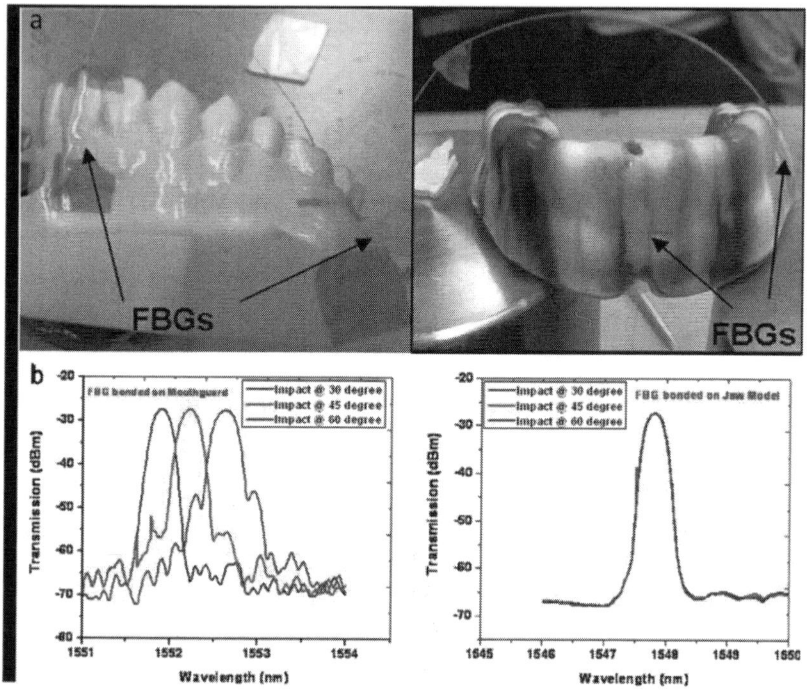

Figure 10. (**a**) The jaw model and mouthguard with FBGs. (**b**) The response of ball impact portraying no changes for FBG bonded on jaw model (**b right**) as compared to FBG bonded on the mouthguard (**b left**) [3,81].

It is no longer strange to combine electronics and textiles into a single structure to produce electronic textiles (e-textiles) [86]. Recently, a new project supported by the Europe Union and called "Optical Fibre Sensors Embedded into technical Textile for Healthcare (OFSETH)" was launched [87]. The aim of OFSETH is to take advantage of pure optical sensing technologies for extending the capabilities of medical technical textiles for wearable health monitoring [88]. These wearable smart textiles can be used to simultaneously monitor the vital physiological parameters such as respiratory movements, cardiac activity, pulse oximetry, and temperature of the body. They exhibit advantageous properties over other standard (non-optical) monitoring techniques, especially for monitoring the anesthetized patients under Medical Resonance Imaging (MRI) [89–91].

The OFSETH project proposed three optical sensing techniques for monitoring the elongation in textile fabrics worn by patients: the bending loss sensor, the FBG sensor and the optical time domain reflectometry sensor. Grillet et al. [87] have reported on these three different techniques, all smart textiles embedded. In this study, only the elongation of the abdominal circumstances during ventilation was measured unlike the study by Wehrle et al. [85] which measured the thorax movement instead. The textile was fixed to a load cell and a movable platform by clamping blocks to investigate the stretching capabilities of the smart textile and the effective strain on the optical fiber was then monitored during the stretching experiment by continuously recording the shifts in Bragg wavelength. The FBG sensor performance during cyclic strains mimicking those normally induced by chest movements was also tested to assess the suitability of this proposed design. The results showed that these three techniques are able to sense the textile elongations between 0% and 3% and the stretching properties of the textile was maintained during the measurement to ensure the patient comfort. Moreover, the FBG sensor embedded in the textile is able to sense weak breathing movements with a better resolution.

Another work by Jonckheere et al. [92] reported on only two pure optical sensing techniques—macro-bending sensor and FBG sensor—embedded into textile fabrics to measure the elongation of the technical textile caused by both the thoracic and abdominal movements during ventilation. The aim is to ensure the proper position of the optical sensors in order for minimizing parasitic effects such as mechanical strains and optical losses, and to convert all the textile elongation into a mechanical force that will be transferred to the sensors. The typical positions of these two sensors are shown in Figure 11(a). The combination of these two designs produces a device allowing for continuous monitoring of both thoracic and abdominal movements. The two designs showed their capability to measure the textile elongation between 0% to 3%. The thoracic and abdominal movements signals obtained (Figure 11(b)) can indicate the respiratory system efficiency and breathing frequency spectrum. This shows that optical sensors can be the golden solution for the measurement of vital signs of patients during MRI examination as it shows immunity to electromagnetic radiations. However, these sensors were then developed to be able to measure both the thoracic and abdominal movements with increased elongation between up to 5% and to be connected to a hard monitor or to a mobile device [93].

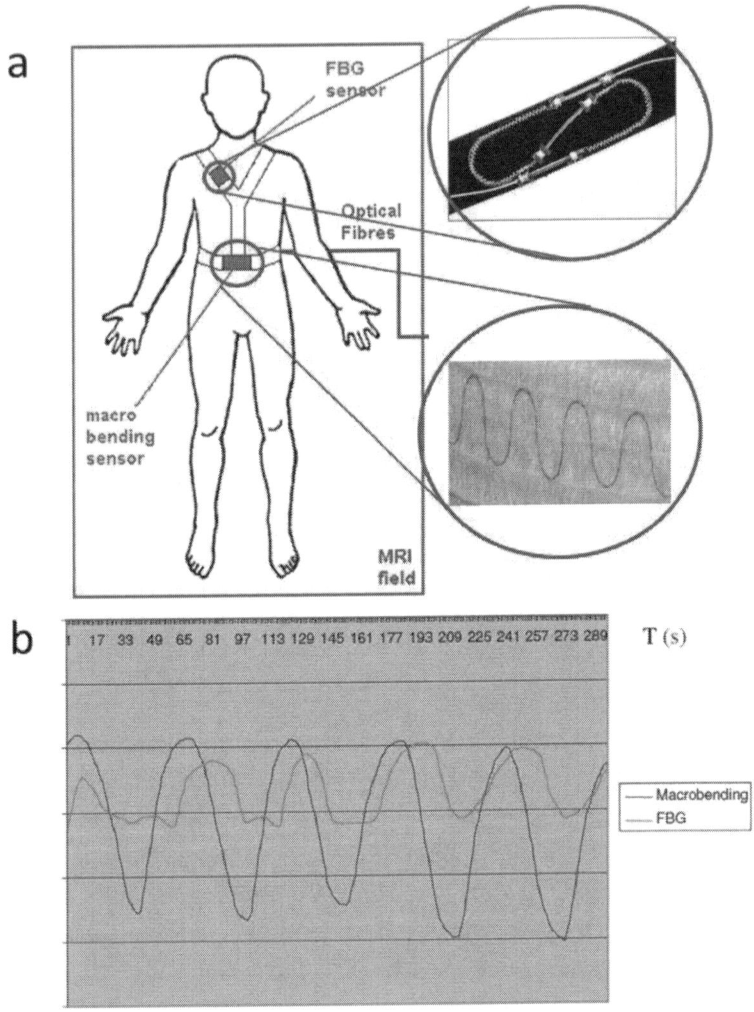

Figure 11. (**a**) The optical fibers-textile integration, illustrating the optimal positions of macro-bending and FBG sensors. (**b**) The signals of both the abdominal and thoracic movements using both types of sensors [92,93].

The FBG sensors previously developed to measure the chest strain due to breathing has been continuously advanced, allowing for detecting both respiratory motions and cardiac frequency simultaneously by using a single FBG, rather than using the macro-bending loss effect of an optical fiber [94–

96]. Although FBG sensors have been used to detect heart sound to evaluate the heart muscles function, which reflect the biomechanics of the heart and great vessels effectively, it was based on an approach enabling the heartbeat measurement only [97]. However, a developed system compatible with different people to measure both components, the respiratory motions and heart rate, using a single FBG sensor was proposed, introducing different measurement methods and sites [98–100]. Silva et al. [98] have developed a FBG sensor embedded in a PVC foil to enable a large sensitive area and enhance the sensitivity. This sensor can be attached to and removed from the chest site via Velcro or even be worn by the subject. A signal processing system was used to filter out the cardiac and respiratory frequencies. A commercial device was utilized for referencing and the optical fibre system was tested on 12 healthy subjects who participated in this study. This system was reliable and found an application niche in MRI environment, where no other types of sensors than optical ones can be used due to EMI.

Wireless technology is widely used in biomedical and life monitoring, such as the monitoring of breath motion. A great effort has been conducted throughout the World to develop mobile approaches for chest movements monitoring. The use of optical fibers embedded into textiles for mobile respiratory motion detection was first reported by D'Angelo et al. [101]. This mobile system involves a T-shirt with an integrated fibre sensor and a portable monitoring unit, and wirelessly connected to a PC to enable the data analysis and visualization. The authors used the bending of the optical fibers to recognize and analyse the respiration of the wearer's chest. However, a novel FBG with a wireless-based approach was introduced by Xiaobin et al. [102]. In this study, the principle of "Four States Method with Combined Modulation" was used in the FBG demodulation, so that this new method could have the features of small size, high integrality and portability. This proposed system can monitor the respiratory movements with a sensitivity of about 41 $\mu\varepsilon$. Furthermore, an alarm signal can be transmitted wirelessly to a remote server when abnormal event occurs. These wireless approaches satisfy the requirements to measure chest deformation during breathing, and, at the same time, provide the convenience to the patients.

3.7. Human Machine Interfaces

Investigation of the pressure ulcer in the Human Machine Interfaces (HMI) is critical to understand the mechanism of the body movement. An embedded array of FBG sensors can be employed in diabetic foot/shoe interface, clinical beds, prosthetic sockets and wheelchair seating systems, targeting pressure ulcer development and wound treatment [103]. Recently, the IASiS project was established, aiming at bringing optical fiber sensing in rehabilitation engineering, especially at the area of HMI research [104,105].

3.7.1. Insole Sensors

Properly-designed temperature-independent FBG sensors can be used to map the foot pressure distribution of diabetic patient's plantar surface through the gait cycle, so as to understand the gait mechanism and diagnose the foot pathology. This helps orthopedic surgeons and doctors determine the actual normal forces exerted at the key pressure points under the patients' foot, the total applied force and the center of pressure. Conventional insole pressure measurement systems employ capacitive [106], conductive [107], piezoelectric [108], resistive [109], and magneto-resistive [110,111] types of transducers which show limited performance and require a heavy shielding, causing an increase in the cost and size of the sensors significantly [112]. The fibre-optic bend loss sensors were also investigated [113], showing good reliability and acceptable accuracy. In contrast to the above systems, FBG pressure sensors employ an all-passive dielectric approach offering many advantageous properties such as electrical isolation from the patients, immunity to EMI, light weight and compactness [4]. In addition, they exhibit higher sensitivity and the capability to be multiplexed and integrated.

A research group based in Singapore has investigated the potential use of FBG-based sensors for monitoring the foot pressure of diabetic patients [112]. Five FBGs with five different wavelengths were fabricated and spatially placed along a single strand of grating fiber in order to monitor different pressure points under foot. They were sandwiched with eight layers of carbon/epoxy laminates on top and two layers at the bottom to form the foot pressure sensor which can be cut into the shape of a footpad. Clinical tests were conducted on a 26-year-old male patient with a body weight of 60 kg. A comparison between the pressure distribution of a normal standing gait and an abnormal standing gait on the right foot were conducted. This sensor shows a better performance than the conventional foot pressure sensing systems in terms of sensitivity, cost and compactness.

Other researchers have also used FBG foot sensors to study the biomechanics of human balancing on a vibrating floor [114]. Three multiplexed FBG sensors were used at the bottom of each foot to provide information on areas where the pressure would be most concentrated. The pressure distribution over different parts of the foot was obtained at different frequencies and different distances from the vibrating source. All observations showed that most of movements occur on the right foot especially around the heel area, which—in turn—show that the right foot is more sensitive to vibration which is consistent with the natural behavior of a normal person whose one leg is more agile and sensitive than the other. These findings show great potential for the application of FBG foot sensors to study the reaction of the human being subjected to floor vibration. Furthermore, the system shows its robustness and ability to work under harsh conditions.

FBG sensors also hold great potential for measuring the in-shoe shear stresses. Although several methods for the shear measurements had been developed [110,111,115,116], they showed common drawbacks which are the relatively large size of the sensors and difficulty in interrogating many sensors at a time to achieve quasi-distributed sensing. In contrast, FBG sensing is attractive method due to its small size and multiplexing capability. Koulaxouzidis et al. [117] investigated the feasibility of FBG sensors to simultaneously measure the in-shoe normal and shear stresses to understand the planter ulcer development after the onset of diabetes. With special configuration, three FBGs were embedded in a block of an epoxy elastomer to be capable of measuring the vertical stresses as well as the magnitude and direction of the shear stresses on its top surface. The experimental results showed the applicability of this design and the effectiveness of the readout system. The good repeatability and high resolution in both directions and its small size are key advantages as well.

3.7.2. Amputee Sockets

There is a potential use of 2D or 3D FBG sensor structures to measure the pressure and shear stresses in the interface between amputee's socket and stump. Since the mid-1960s, researchers have proposed a variety of measurement systems such as ESG based transducers [118–120], F-socket transducer arrays [121], and FEA [122–124] in order to understand the pressure distribution in this interface, leading to fabrication of amputee sockets that prevent discomfort and provide satisfactory results. Despite these technological advances in the existing socket design and the measurement systems, many amputees still complain that their prosthesis causes complications such as edema, pressure ulcers, skin irritation and dermatitis [125]. However, the use of FBG sensors could show more accurate results which can assist in the development of better socket designs that help amputees live normal day-to-day lives with minimal complications. Kanellos et al. [126] developed a flexible 2D FBG pressure sensing sheet suitable for biomechanical applications such as amputee socket/stump interface. The sensor mechanical properties matched human skin behaviour and its operational performance exhibited a maximum fractional pressure sensitivity of 12 MPa-1 with a spatial resolution of 1 × 1 cm^2 and demonstrated no hysteresis and real-time operation. These attractive properties meet the requirements of HMI pressure measurements including amputee sockets.

In the EC FP7 project "IASiS" studies have introduced an FBG based 2D sensing surface that could continuously monitor the pressure and potentially shear applied to below-knee stump/socket interface [127,128]. Pleros et al. [103] also demonstrated the evolutionary road towards smart FBG-based HMI structures for pressure ulcer monitoring in amputee socket HMI systems. They addressed the pressure and shear by using arrays of multiplexed FBG sensors by introducing two parallel layers of 2D sensing FBG meshes allowing for shear sensing by correlating the differences in the strain-induced deformation of the two FBG mesh layers. These two mesh layers form a 3D sensor structure that could measure both the pressure and shear stress at a time. This developed sensor is suited to be utilized in HMIs structures such as stump/socket interfaces.

A recent study by Kanellos, Tsiokos, Pleros, Childs and Pissadakis [105] presented the optimal design of FBG sensing pad that exhibits enhanced durability and higher sensitivity and is suitable for applications in HMI surfaces such as amputee sockets, medical beds and wheelchairs. They investigated three case studies: fibre embodiment depth (centre and top position of pad cross-section), pad thickness (2 and 3 mm) and fibre type (hydrogenated SMF-28 and non-hydrogenated GF1B). In the first study case, two hydrogenated SMF-28 fibre FBG sensors were embedded in a 2 mm thick PDMS sheet in the centre and top position of the cross-section. A series of vertical loads were applied using a gauge test stage. The results revealed a 50% increased sensitivity and a 40% increased durability for the FBG positioned in the center of the PDMS layer. In the second case study, they compared the sensitivity and durability of hydrogenated SMF-28 fibre FBG sensors embedded in the center of a 2 mm and 3 mm thick PDMS sheets respectively. After the test, it was found out that the sensor sensitivity was increased with 50% and a marginal increase in the durability for the thicker (3 mm) sensor pad. Finally they compared the sensitivity of two FBGs inscribed in two different fiber types (hydrogenated SMF-28 and non-hydrogenated GF1B). The results exhibited 20% increased sensitivity for the hydrogenated fiber and a 40% decrease in its dynamic range when compared to the non-hydrogenated fiber. It is clear that the optimum sensor pad layout enhancing durability and preserving required sensitivity and can be successfully implemented in HMI applications is when the FBG is written in a non-hydrogenated fiber and embedded at the centre of the thicker (3 mm) PDMS sheet.

The previous researches proposed sensors that could be suitable for simultaneous measurement of the normal and shear stresses at the amputee socket/stump interfaces. To date, full-scale implementation of these sensors in this area of research has not been established yet. In a project entitled "Synthetic Prosthetic Socket through Stump—Liner Interfacial Stresses Measurement" which supported by the Ministry of Higher Education—Malaysia [129], our research group and in collaboration with Photonics Research Group [130] in the department of electrical engineering at University of Malaya aims to establish investigative studies on the capability of the optical FBG sensors to measure socket/stump interface pressures, leading to the development of a novel interfacial stresses measuring system that can also be applied over surfaces of irregular shapes. Currently, we are focusing on designing an FBG sensor that can be used in transtibial amputee socket/stump interfaces and capable to measure the normal and shear stresses simultaneously. We started to implement a single FBG embedded in a polymeric material and bonded at the patellar tendon bar area in order to measure the pressure at this most tolerant area within the amputee socket. This sensor was tested in real-time by inserting a heavy duty balloon into the socket and inflating it via an air compressor in order for simulating the pressure applied by real stump and for examining the sensor sensitivity and repeatability in-situ (Figure 12). The results exhibited that the sensor has a good sensitivity and acceptable hysteresis. We are currently working hard to involve subjects in our experiments in order to clinically validate this design. Plans to implement experiments to map the pressure as well as the shear stress over the whole inner socket wall and compare this system to the conventional ones are currently in progress.

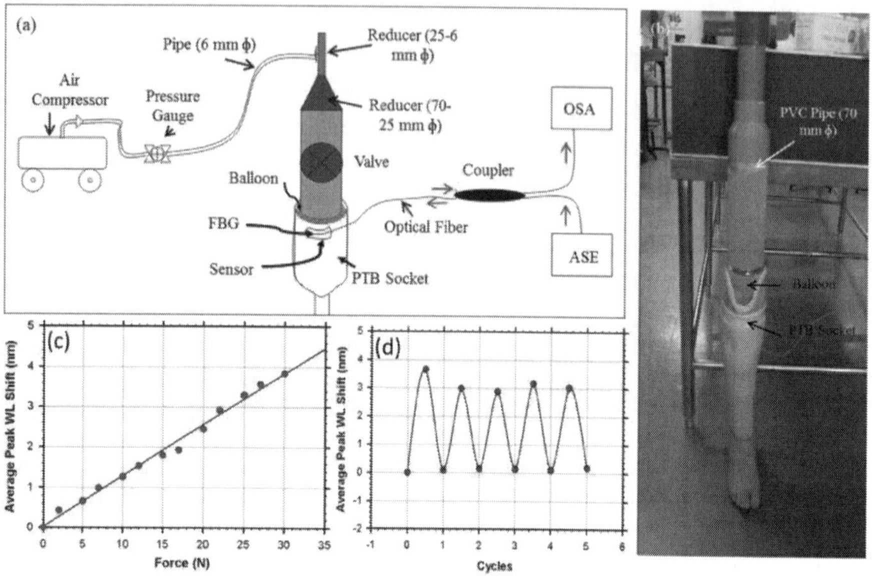

Figure 12. (**a**) A simple schematic diagram of the in situ test. It shows the sensor while connected to OSA and subjected to loads. (**b**) Depicts the image of the set up. (**c**) The calibration of FBG sensor showing the FBG peak wavelength shift versus the applied force. (**d**) Cyclic change in pressure from minimum (0 Pa) to 35 kPa was performed to test the repeatability and reliability of the sensor. The solid line is for indicative purpose only.

3.7.3. Bedridden Patients

The utilization of FBG sensors in HMI systems was extended to their use in hospitals and nursing homes to automatically monitor the behaviour of clinical bedridden patients and their vital signs. When paralysed patients lying on beds stay unmoved for long time, bedsores generation occurs. Currently, monitoring of bedridden patients is usually handled by caregivers and professionals who must manually conduct periodic checks. This practice is tedious and time consuming and the occurrence of delayed response is highly expected. Therefore, development of multiplexed FBG pressure sensors which can be strategically distributed on the top surface of clinical beds is highly recommended to continuously monitor the patients sleeping and sitting pattern and to alert the nurse to help move the patient's body when unmoved for long time, in order to prevent bedsore generation.

Hao et al. [131] have developed a smart multi-functional FBG sensor system which can map out the pressure points beneath the patients and monitor the respiratory rates and other parameters as well. In this study, twelve FBG sensors embedded in arc-shaped carbon fibre reinforced plastic material were cascaded along a single fibre and mounted on the surface of the bed to form 3 × 4 matrix array, and the bed was then covered by the usual mattress (Figure 13). A sample group of 10 subjects were undergone trials in the laboratory and the system captured and displayed the body movement and respiration pattern. This system showed promising effectiveness and ability to work. Such systems show advantageous features if compared with the current commercial systems demanding the patients to wear specialized probes which introduce some level of discomfort. However, more studies are recommended to advance these systems to be able to accurately capture most of the vital signs that help practitioners observe the bedridden patients properly.

3.8. Tendons and Ligaments

Accurate and precise measurements of ligament and tendon biomechanics in living humans are vitally essential in better understanding function and injury and in optimizing treatment [132]. FBG sensors have been used to study the biomechanical properties of tendons and ligaments and shown encouraging results. Vilimek proposed the use of FBG sensors to investigate the in vitro biomechanical properties of porcine tendons using a tensile machine (MTS Minibionix) in vertical orientation [133]. The FBG sensor was crosswise included by grouting point into the tendon whose ends were attached to the actuator using the freeze clamping technique. The sensor was repeatedly loaded and unloaded in steps using the MTS and the measurement was run until the tendon was destructed. In all of the trials, the measured forces by MTS and measured signals from FBG sensor give the same proportional changes and it is possible to estimate the loading force. The FBG sensors were more sensitive and accurate than other fibre-optic methods based on decrement of light properties during pressing change of cross section size. This approach is suitable for in vivo measurement of a musculotendon forces in humans for validation the muscle force computational models because the injury after application is minor.

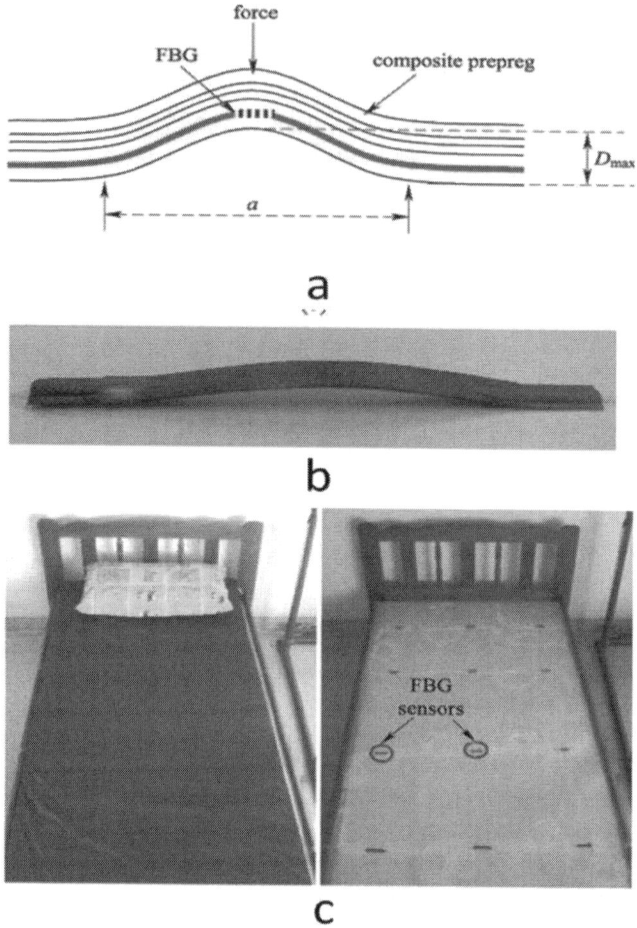

Figure 13. (**a**) A drawing illustrating the FBG embedded in a composite laminates. (**b**) Arc-shaped sensor. (**c**) A 3 × 4 array of FBGs deployed on bed [131].

Ren et al. proposed a FBG-based displacement sensor for monitoring strains developed in tendons and ligaments in different postures and in locomotion, and then compared the obtained results to those obtained using traditional camera displacement sensors [134]. The FBG was embedded in a micro-shape-memory-alloy tube which was then placed on an Achilles tendon surface extracted from human cadaver and the whole system was mounted on a material testing device for loading test. After sensor calibration in the laboratory, the performance of the FBG sensors and traditional camera displacement sensors was compared. Additional experiments were performed in cadaveric knees to assess the feasibility of the FBG sensors to measure ligament deformation in a variety of simulated postures. The results demonstrate that the proposed FBG sensor is a highly accurate, easily implantable, and minimally invasive method of measuring tendon and ligament displacement as compared to the traditional measurement methods.

3.9. Human Body kinematics

The monitoring of different human body kinematics and the analysis of posture and gesture are crucial in bioengineering and rehabilitation. It aims at improving athletic performance in competitions and at evaluating frequently the patients undergoing therapeutic treatments to know the efficiency of the prescribed therapy. Several systems have been previously proposed for monitoring human body kinematics [135–139], however, they all are not able to withstand some specific environments like chemical solutions, electromagnetic noise and high temperature, among other [140].

Recently, a research group based in Portugal has proposed a smart skin polymeric foil based on FBG sensors which is suitable for the measurement of body kinematics [141–146]. Rocha et al. [147] have validated this wearable system by monitoring maximum knee flexion and extension and all the movements in-between during a full period of human gait. A single FBG was placed at the center of the knee joint and embedded in a rectangular-shaped structure that is composed by three layers of flexible and stretchable PVC materials. This sensing structure can be easily attached to/removed from an elastic knee band by using small metallic pressure buttons to ensure that the FBG is capable of detecting only the flexion and extension as the subject moves around. The subject underwent different types of running and walking on a commercially-available treadmill in order to assess the reliability and consistency of this system and the movements of knee joints were being measured as a function of the FBG wavelength, and, at the same time, with video recording for comparison purposes. The results show that the stage where the maximum flexion/extension in the knee joint, and also on the FBG, represents the maximum/minimum wavelength values as depicted in Figure 14(a,b). This sensing structure shows a good sensitivity to accurately measure the knee flexion/extension during the walking and running tests and is able to be used for any joint in human body. This FBG-based PVC foil opens up the road for many applications in biomechanics and other arenas.

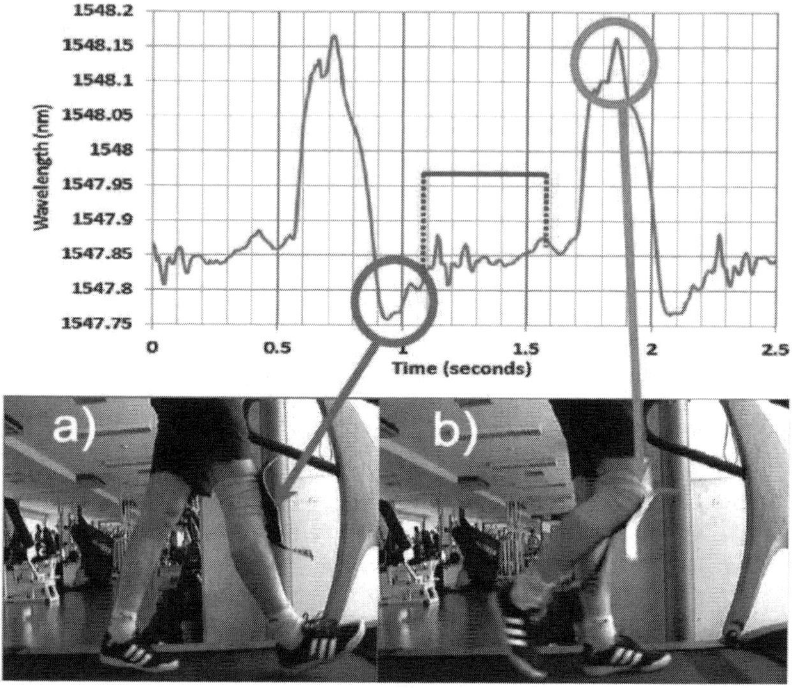

Figure 14. The wavelength values for a walking at 4 km/h. (**a**) and (**b**) show the minimum and maximum deflections of the FBG, respectively [147].

The monitoring of hand gesture and posture of brain stroke patients is of interest in rehabilitation engineering in order to assess the hand ability to perform action in a proper manner during therapeutic sessions. Several wearable glove systems using different approaches have been proposed [148–152], but demonstrated some concerns [153]. Silva et al. have designed a wearable sensing glove based on FBGs in order to enable the measurement of angles between the finger phalanxes and to overcome the limitations presented by other previously reported methods [153]. Fourteen FBGs accommodated in a single optical fiber were strategically placed in a curvilinear layout and embedded in the middle layer of a flexible PVC foil cut into the hand shape so that each FBG would be positioned over each of the 14 phalanx joints present in a single hand. Thereafter this customized hand-shaped sensing foil was sewed to the upper face of a standard fabric glove (Figure15(a)) to guarantee that every FBG is positioned in its intended place throughout the experiments. Tests to evaluate the FBGs performance and functionality according to hand flexion and extension were performed. The sensor exhibited a sinusoid-like response (Figure 15(b), left) where the positive and negative limits represent the hand closing and opening actions, respectively. It is possible to estimate the strength from which the 1 nm wavelength shift resulted from an applied load of 7.8 N, leading to retrieve information about joint angles. A comparison between the real and measured angles produced a linear fit, representing the system accuracy with R-square of 0.99929 and a slope of 0.99869 (Figure 15(b), right). In the second part of this study, a hand motion capture system based in a 3-D virtual model of the hand was built (Figure15(a)), allowing for the visualization of the hand movement in real-time on a PC which can be used to provide information about the hand angles, strength, and movement range. This wearable system could have a great potential in physical therapy applications, in particular for hand-impaired people, and when studying human kinematics, among others.

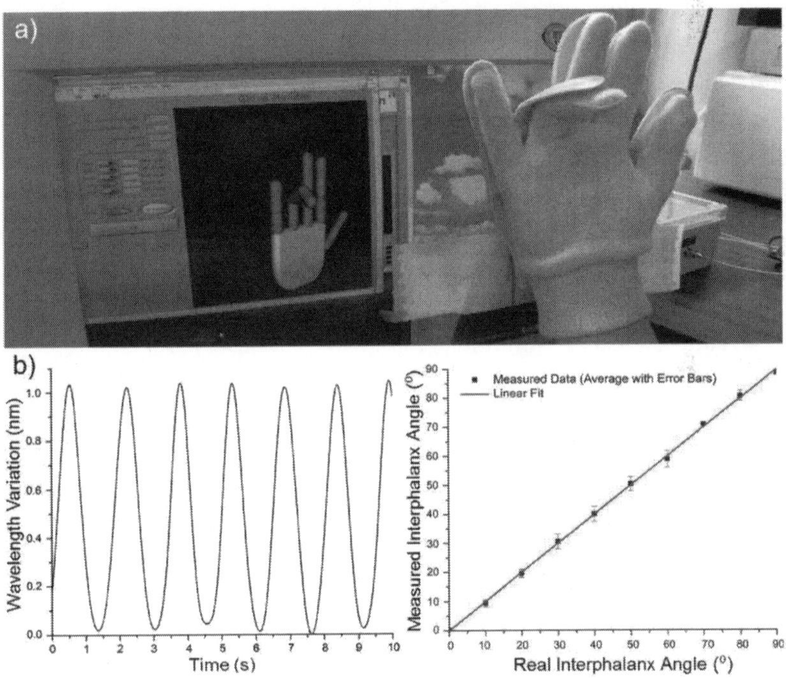

Figure 15. (**a**) A hand-shaped PVC foil accommodating 14 FBGS and sewed to the upper face of a standard glove, allowing for the visualization of the hand movement in real-time on a PC. (**b**) (left) The response of ring finger FBG sensor for opening and closing hand movements. (right) A curve comparing between real and measured angles, illustrating the system accuracy [153].

4. CONCLUSIONS

It is clearly evident from the literature that FBGs can be successfully employed in a wide range of applications in biomechanics and rehabilitation engineering. The reason of this is because FBGs have a number of distinguishing advantages, over other traditional systems, such as their small size, chemical inertness, light weight, biocompatibility and multiplexing capability. Although FBGs have proven their feasibility and superiority over existing technologies, they have not been commercialized yet in the field of biomechanics and rehabilitation. With the rapid developments in this field and the need to successfully implement FBG sensors, a collaborative work involving medical doctors, experts in fiber-optics and programming engineers must take place in order to establish the practicality of this technology, taking into consideration that FBG sensors must be user-friendly, cheap in cost, and easy-to-use products so as to satisfy the requirements of medical doctors and end users as well. Very recently, portable interrogation systems with high-speed data acquisition (up to 5,000 Hz) were made available, leading to broadened research areas and finding new niche applications. This article is a comprehensive review that covers all the possible uses of FBG sensors and the ongoing researches worldwide in biomechanics and rehabilitation engineering. It also exposes the superior advantages of FBGs when compared to the other existing technologies. It is believed that FBG sensors will undoubtedly have a major role in the future and provide effective solutions for a wide variety of applications in biomechanics and rehabilitation engineering.

ACKNOWLEDGMENTS

This work was jointly supported by UM/MOHE/HIR Grants, Projects No. (D000014-16001) and (A000007-50001).

REFERENCES

1. Dennison, C.R.; Wild, P.M.; Wilson, D.R.; Cripton, P.A. A minimally invasive in-fiber Bragg grating sensor for intervertebral disc pressure measurements. Meas. Sci. Technol. 2008, 19, 085201.

2. Mihailov, S.J. Fiber Bragg grating sensors for harsh environments. Sensors 2012, 12, 1898–1918.

3. Mishra, V.; Singh, N.; Tiwari, U.; Kapur, P. Fiber grating sensors in medicine: Current and emerging applications. Sensor. Actuator. A Phys. 2011, 167, 279–290.

4. Spillman, W.B., Jr. Fiber Optic Sensors: An Introduction for Engineers and Scientists, 2nd ed.; Wiley: New York, NY, USA, 2011.

5. Kersey, A.D.; Davis, M.A.; Patrick, H.J.; LeBlanc, M.; Koo, K.; Askins, C.; Putnam, M.; Friebele, E.J. Fiber grating sensors. J. Lightwave Technol. 1997, 15, 1442–1463.

6. Cheng, S.; Zhang, Y. Application of fiber grating sensor for oil leak detection in oil industry. Appl. Opt. 2008, 29, 441–443.

7. Hassan, M.; Tamchek, N.; Abas, A.; Johar, R.; Adikan, F. Dual-phase sensing for early detection of prepreg structural failures via etched cladding Bragg grating. Sensor. Actuator. A Phys. 2011, 171, 126–130.

8. Schizas, C.; Stutz, S.; Botsis, J.; Coric, D. Monitoring of non-homogeneous strains in composites with embedded wavelength multiplexed fiber Bragg gratings: A methodological study. Compos. Struct. 2012, 94, 987–994.

9. Zhao, Y.; Zhao, Y.; Zhao, M. Novel force sensor based on a couple of fiber Bragg gratings. Measurement 2005, 38, 30–33.

10. Urban, F.; Kadlec, J.; Vlach, R.; Kuchta, R. Design of a pressure sensor based on optical fiber bragg grating lateral deformation. Sensors 2010, 10, 11212–11225.

11. Ng, J.H.; Zhou, X.; Yang, X.; Hao, J. A simple temperature-insensitive fiber Bragg grating displacement sensor. Opt. Commun. 2007, 273, 398–401.

12. Zhan, Y.; Xue, S.; Yang, Q.; Xiang, S.; He, H.; Zhu, R. A novel fiber Bragg grating high-temperature sensor. Optik 2008, 119, 535–539.

13. Yeo, T.L.; Sun, T.; Grattan, K.T.V.; Parry, D.; Lade, R.; Powell, B.D. Characterisation of a polymer-coated fibre Bragg grating sensor for relative humidity sensing. Sensor. Actuator. B Chem. 2005, 110, 148–156.

14. Fernandez Fernandez, A.; Brichard, B.; Berghmans, F.; Decreton, M. Dose-rate dependencies in gamma-irradiated fiber Bragg grating filters. IEEE Trans. Nucl. Sci. 2002, 49, 2874–2878.

15. Kalinowski, H.J. Fibre Bragg grating applications in Biomechanics. Proc. SPIE 2008, 7004, 700430.

16. Yan, C.; Ferraris, E.; Reynaerts, D. A pressure sensing sheet based on optical fibre technology. Procedia Eng. 2011, 25, 495–498.

17. Luyckx, G.; Voet, E.; Lammens, N.; Degrieck, J. Strain measurements of composite laminates with embedded fibre Bragg gratings: Criticism and opportunities for research. Sensors 2011, 11, 384–408.

18. Yan, C.; Ferraris, E.; Geernaert, T.; Berghmans, F.; Reynaerts, D. Development of flexible pressure sensing polymer foils based on embedded fibre Bragg grating sensors. Procedia Eng. 2010, 5, 272–275.

19. Chen, Y.C.; Hsieh, C.C.; Lin, C.C. Strain measurement for composite tubes using embedded, fiber Bragg grating sensor. Sensor. Actuator. A Phys. 2011, 167, 63–69.

20. Frieden, J.; Cugnoni, J.; Botsis, J.; Gmür, T.; Coric, D. High-speed internal strain measurements in composite structures under dynamic load using embedded FBG sensors. Compos. Struct. 2010, 92, 1905–1912.

21. Kashyap, R. Fiber Bragg Gratings., 2nd ed.; Academic Press: Burlington, MA, USA, 2010.

22. Othonos, A.; Kalli, K. Fiber Bragg Gratings: Fundamentals and Applications in Telecommunications and Sensing; Artech House: Norwood, MA, USA, 1999.

23. Rodrigues, C.; Félix, C.; Lage, A.; Figueiras, J. Development of a long-term monitoring system based on FBG sensors applied to concrete bridges. Eng. Struct. 2010, 32, 1993–2002.

24. Tjin, S.; Suresh, R.; Ngo, N. Fiber Bragg grating based shear-force sensor: Modeling and testing. J. Lightwave Technol. 2004, 22, 1728–1733.

25. Zhang, W.; Li, F.; Liu, Y. FBG pressure sensor based on the double shell cylinder with temperature compensation. Measurement 2009, 42, 408–411.

26. Fresvig, T.; Ludvigsen, P.; Steen, H.; Reikerås, O. Fibre optic Bragg grating sensors: An alternative method to strain gauges for measuring deformation in bone. Med. Eng. Phys. 2008, 30, 104–108.

27. Milgrom, C.; Finestone, A.; Segev, S.; Olin, C.; Arndt, T.; Ekenman, I. Are overground or treadmill runners more likely to sustain tibial stress fracture? Brit. J. Sport. Med. 2003, 37, 160–163.

28. Milgrom, C.; Finestone, A.; Simkin, A.; Ekenman, I.; Mendelson, S.; Millgram, M.; Nyska, M.; Larsson, E.; Burr, D. In vivo strain measurements to evaluate the strengthening potential of exercises on the tibial bone. J. Bone Joint Surg. Brit.Vol. 2000, 82, 591–594.

29. Carlisle, E.M. Silicon: A possible factor in bone calcification. Science 1970, 167, 279–280.

30. Talaia, P.; Ramos, A.; Abe, I.; Schiller, M.; Lopes, P.; Nogueira, R.; Pinto, J.; Claramunt, R.; Simões, J. Plated and intact femur strains in fracture fixation using fiber Bragg gratings and strain gauges. Exp. Mech. 2007, 47, 355–363.

31. Mishra, V.; Singh, N.; Rai, D.; Tiwari, U.; Poddar, G.; Jain, S.; Mondal, S.; Kapur, P. Fiber Bragg grating sensor for monitoring bone decalcification. Orthop. Traumatol. Sur. 2010, 96, 646–651.

32. Carvalho, L.; Alberto, N.J.; Gomes, P.S.; Nogueira, R.N.; Pinto, J.L.; Fernandes, M.H. In the trail of a new bio-sensor for measuring strain in bone: Osteoblastic biocompatibility. Biosens. Bioelectron. 2011, 26, 4046–4052.

33. Reikeras, O.; Aarnes, G.T.; Steen, H.; Ludvigsen, P.; Sagvolden, G. Differences in external and internal cortical strain with prosthesis in the femur. Open J. Orthopedics 2011, 5, 379–384.

34. Completo, A.; Fonseca, F.; Relvas, C.; Ramos, A.; Simões, J. Improved stability with intramedullary stem after anterior femoral notching in total knee arthroplasty. Knee Surg. Sports Traumatol. Arthrosc. 2012, 20, 487–494.

35. Hirsh, D.; Bhalla, S.; Roffman, M. Supracondylar fracture of the femur following total knee replacement. Report of four cases. J. Bone Joint Surg. Amer.Vol. 1981, 63, 162–163.

36. Mann, K.A.; Gupta, S.; Race, A.; Miller, M.A.; Cleary, R.J.; Ayers, D.C. Cement microcracks in thin-mantle regions after in vitro fatigue loading. J. Arthroplasty 2004, 19, 605–612.

37. Stolk, J.; Verdonschot, N.; Cristofolini, L.; Toni, A.; Huiskes, R. Finite element and experimental models of cemented hip joint reconstructions can produce similar bone and cement strains in pre-clinical tests. J. Biomech. 2002, 3, 499–510.

38. Ramos, A.; Abe, I.; Schiller, M.; Lopes, P.; Nogueira, R.; Pinto, J.; Simões, J. On the use of fiber bragg sensors to assess temperature and thermal induce strain profiles in cemented hip mantles. J. Biomech. 2006, 39, S514.

39. Frias, C.; Frazão, O.; Tavares, S.; Vieira, A.; Marques, A.; Simões, J. Mechanical characterization of bone cement using fiber Bragg grating sensors. Mater. Design 2009, 30, 1841–1844.

40. Harris, M.; Morberg, P.; Bruce, W.; Walsh, W. An improved method for measuring tibiofemoral contact areas in total knee arthroplasty: a comparison of K-scan sensor and Fuji film. J. Biomech. 1999, 32, 951–958.

41. Kdolsky, R.; Arabid, B.; Fuchs, M.; Schabus, R.; Vécsei, V. Measuring the pressure pattern of the joint surface in the uninjured knee. Wien. Klin. Wochenschr. 2004, 116, 196–200.

42. D'Lima, D.D.; Townsend, C.P.; Arms, S.W.; Morris, B.A.; Colwell, C.W., Jr. An implantable telemetry device to measure intra-articular tibial forces. J. Biomech. 2005, 38, 299–304.

43. Drewniak, E.I.; Crisco, J.J.; Spenciner, D.B.; Fleming, B.C. Accuracy of circular contact area measurements with thin-film pressure sensors. J. Biomech. 2007, 40, 2569–2572.

44. Mohanty, L.; Tjin, S.C. Pressure mapping at orthopaedic joint interfaces with fiber Bragg gratings. Appl. Phys. Lett. 2006, 88, 083901.

45. Mohanty, L.; Tjin, S.C.; Lie, D.T.T.; Panganiban, S.E.C.; Chow, P.K.H. Fiber grating sensor for pressure mapping during total knee arthroplasty. Sensor. Actuator. A Phys. 2007, 135, 323–328.

46. Anderson, A.E.; Ellis, B.J.; Maas, S.A.; Peters, C.L.; Weiss, J.A. Validation of finite element predictions of cartilage contact pressure in the human hip joint. J. Biomech. Eng. 2008, 130, 051008.

47. Wilson, D.R.; Mc Walter, E.J.; Johnston, J.D. The measurement of joint mechanics and their role in osteoarthritis genesis and progression. Med. Clin. N. Amer. 2009, 93, 67–82.

48. Dennison, C.R.; Wild, P.M.; Wilson, D.R.; Gilbart, M.K. An in-fiber Bragg grating sensor for contact force and stress measurements in articular joints. Meas. Sci. Technol. 2010, 21, 115803.

49. Nachemson, A. The load on lumbar disks in different positions of the body. Clin. Orthop. Related Res. 1966, 45, 107–122.

50. Nachemson, A.; Morris, J.M. In vivo measurements of intradiscal pressure discometry, a method for the determination of pressure in the lower lumbar discs. J. Bone Joint Surg. Amer.Vol. 1964, 46, 1077–1092.

51. Cunningham, B.W.; Kotani, Y.; McNulty, P.S.; Cappuccino, A.; McAfee, P.C. The effect of spinal destabilization and instrumentation on lumbar intradiscal pressure: An in vitro biomechanical analysis. Spine 1997, 22, 2655–2663.

52. Nachemson, A. Lumbar intradiscal pressure. Experimental studies on post-mortem material. Acta Orthop. Scand. 1960, 43, 1–104.

53. Wilke, H.J.; Wolf, S.; Claes, L.E.; Arand, M.; Wiesend, A. Influence of varying muscle forces on lumbar intradiscal pressure: An in vitro study. J. Biomechan. 1996, 29, 549–555.

54. McNally, D.; Adams, M. Internal intervertebral disc mechanics as revealed by stress profilometry. Spine 1992, 17, 66–73.

55. Wilke, H.J.; Neef, P.; Caimi, M.; Hoogland, T.; Claes, L.E. New in vivo measurements of pressures in the intervertebral disc in daily life. Spine 1999, 24, 755–762.

56. Yantzer, B.K.; Freeman, T.B.; Lee, W.E., III; Nichols, T.; Inamasu, J.; Guiot, B.; Johnson, W.M. Torsion-induced pressure distribution changes in human intervertebral discs: An in vitro study. Spine 2007, 32, 881–884.

57. Pospiech, J.; Stolke, D.; Wilke, H.J.; Claes, L.E. Intradiscal pressure recordings in the cervical spine. Neurosurgery 1999, 44, 379–384.

58. Steffen, T.; Baramki, H.G.; Rubin, R.; Antoniou, J.; Aebi, M. Lumbar intradiscal pressure measured in the anterior and posterolateral annular regions during asymmetrical loading. Clin. Biomech. 1998, 13, 495–505.

59. Hattori, S.; Oda, H.; Kawai, S. Cervical intradiscal pressure in movements and traction of the cervical spine. Z Orthop. 1981, 119, 568–569.

60. Cripton, P.; Dumas, G.; Nolte, L.P. A minimally disruptive technique for measuring intervertebral disc pressure in vitro: Application to the cervical spine. J. Biomech. 2001, 34, 545–549.

61. Seo, K.S.; Derby, R.; Date, E.S.; Lee, S.H.; Kim, B.J.; Lee, C.H. In vitro measurement of pressure differences using manometry at various injection speeds during discography. Spine J. 2007, 7, 68–73.

62. Dennison, C.R.; Wild, P.M.; Dvorak, M.F.; Wilson, D.R.; Cripton, P.A. Validation of a novel minimally invasive intervertebral disc pressure sensor utilizing in-fiber Bragg gratings in a porcine model: An ex vivo study. Spine 2008, 33, E589–E594.

63. Dennison, C.R.; Wild, P.M.; Byrnes, P.W.G.; Saari, A.; Itshayek, E.; Wilson, D.C.; Zhu, Q.A.; Dvorak, M.F.S.; Cripton, P.A.; Wilson, D.R. Ex vivo measurement of lumbar intervertebral disc pressure using fibre-Bragg gratings. J. Biomech. 2008, 41, 221–225.

64. Wild, P.M.; Dennison, C.R. Pressure Sensor for Biological Fluids and Use Thereof. U.S. Patent 070213, 2007.

65. Roriz, P.; Abe, I.; Schiller, M.; Gabriel, J.; Simões, J. Ex vivo intervertebral disc bulging measurement using a fibre bragg grating sensor. Exp. Mech. 2011, 51, 1573–1577.

66. Tjin, S.; Tan, Y.; Yow, M.; Lam, Y.Z.; Hao, J. Recording compliance of dental splint use in obstructive sleep apnoea patients by force and temperature modelling. Med. Biol. Eng. Comput. 2001, 39, 182–184.

67. Ottevaere, H.; Tabak, M.; Aznar, D.; Fernandez, A.; Van Ierschot, S.; Berghmans, F.; Thienpont, H. Optical Fiber Sensors for Monitoring Stress Build-Up in Dental Cements. Proceedings of the 16th International Conference on Optical Fibre Sensors, Nara, Japan, 13–17 October 2003.

68. Ottevaere, H.; Tabak, M.; Fernandez, A.F.; Berghmans, F.; Thienpont, H. Optical fiber sensors and their application in monitoring stress build-up in dental resin cements. Proc. SPIE 2005, 5952, 204–216.

69. Fernandez, A.; Ottevaere, H.; Van Ierschot, C.; Panajatov, K.; Berghmans, F.; Thienpont, H. Multi-Parameter Force Sensing with Fiber Bragg Grating Sensors. Proceedings of the 2002 Symposium of the IEEE/LEOS Benelux Chapter, Amsterdam, The Netherlands, 9 December 2002; pp. 139–144.

70. Milczewski, M.; da Silva, J.; Abe, I.; Carvalho, L.; Nogueira, R.; Paterno, A.; Kalinowski, H.; Pinto, J. Determination of setting expansion of dental materials using fibre optical sensing. Meas. Sci. Technol. 2006, 17, 1152–1156.

71. Carvalho, L.; Vaz, M.; Simões, J. Mandibular strains induced by conventional and novel dental implants. J. Strain Anal. Eng. Design. 2004, 39, 291–297.

72. Carvalho, L.; Vaz, M.; Simões, J. Determination of strains due to impact loads on an implanted cadaveric mandible. Acta Bioeng. Biomed. 2003, 4, 801–802.

73. Kalinowski, H.; Simoes, J.; Pinto, J.; Carvalho, L.; Nogueira, R.; Silva, J.; Milczewski, M. Probing Dental Biomechanics with Fibre Bragg Grating Sensors. Proceedings of the 7th International Conference on Optoelectronics, Fiber Optics & Photonics, Cochin, India, 9–11 December 2004.

74. Silva, J.C.C.; Carvalho, L.; Nogueira, R.N.; Simoes, J.A.; Pinto, J.L.; Kalinowski, H.J. FBG applied in dynamic analysis of an implanted cadaveric mandible. Proc. SPIE 2004, 5502, 226–229.

75. Carvalho, L.; Silva, J.C.C.; Nogueira, R.; Pinto, J.; Kalinowski, H.; Simúes, J. Application of Bragg grating sensors in dental biomechanics. J. Strain Anal. Eng. Design 2006, 41, 411–416.

76. Silva, J.; Ramos, A.; Carvalho, L.; Nogueira, R.; Ballu, A.; Mesnard, M.; Pinto, J.; Kalinowski, H.J.; Simoes, J. Fibre Bragg grating sensing and finite element analysis of the biomechanics of the mandible. Proc. SPIE 2005, 5855, 102–105.

77. Carvalho, P.; Abe, I.; Schiller, M.; Carvalho, L.; Simões, J.; Lopes, P.; Pinto, J. FEA and experimental FBG sensing system for the analysis of different dental implant concepts. J. Biomech. 2006, 39, S568–S568.

78. Schiller, M.W.; Abe, I.; Carvalho, P.; Lopes, P.; Carvalho, L.; Nogueira, R.N.; Pinto, J.L.; Simões, J. A. On the Use of FBG Sensors to Assess the Performance of a Dental Implant System. Proceedings of the 18th International Conference on Optical Fibre Sensors, Cancún, México, 23–27 October 2006.

79. Milczeswki, M.; Silva, J.; Abe, I.; Simões, J.; Paterno, A.; Kalinowski, H. Measuring Orthodontic Forces with HiBi FBG Sensors. Proceedings of the 18th International Conference on Optical Fibre Sensors, Cancún, México, 23–27 October 2006.

80. Milczewski, M.S.; Kalinowski, H.J.; Da Silva, J.C.C.; Abe, I.; Simões, J.A.; Saga, A. Stress monitoring in a maxilla model and dentition. Proc. SPIE 2011, 7753, 77534V.

81. Tiwari, U.; Mishra, V.; Bhalla, A.; Singh, N.; Jain, S.C.; Garg, H.; Raviprakash, S.; Grewal, N.; Kapur, P. Fiber Bragg grating sensor for measurement of impact absorption capability of mouthguards. Dent. Traumatol. 2011, 27, 263–268.

82. Massagram, W.; Lubecke, V.M.; Boric-Lubecke, O. Microwave Non-Invasive Sensing of Respiratory Tidal Volume. Proceedings of the 31th Annual International Conference of the IEEE Engineering in Medicine and Biology Society, Minneapolis, USA, 3– 6 September 2009; pp. 4832–4835.

83. Bai, Y.W.; Li, W.T.; Chen, Y.W. Design and Implementation of an Embedded Monitor System for Detection of a Patient's Breath by Double Webcams. Proceedings of the 2010 IEEE International Workshop on Medical Measurements and Applications, Ottawa, ON, Canada, 30 April–1 May 2010; pp. 171–176.

84. Babchenko, A.; Khanokh, B.; Shomer, Y.; Nitzan, M. Fiber optic sensor for the measurement of the respiratory chest circumference changes. J. Biomed. Opt. 1999, 4, 224–229.

85. Wehrle, G.; Nohama, P.; Kalinowski, H.J.; Torres, P.I.; Valente, L.C.G. A fibre optic Bragg grating strain sensor for monitoring ventilatory movements. Meas. Sci. Technol. 2001, 12, 805–809.

86. Marculescu, D.; Marculescu, R.; Park, S.; Jayaraman, S. Ready to ware. IEEE Spectr. 2003, 40, 28–32.

87. Grillet, A.; Kinet, D.; Witt, J.; Schukar, M.; Krebber, K.; Pirotte, F.; Depré, A. Optical fiber sensors embedded into medical textiles for healthcare monitoring. IEEE Sens. J. 2008, 8, 1215–1222.

88. Optical Fibre Sensors Embedded into Technical Textile for Healthcare. Available online: www.ofseth.org (accessed on 12 July 2012).

89. De jonckheere, J.; Jeanne, M.; Grillet, A.; Weber, S.; Chaud, P.; Logier, R.; Weber, J. OFSETH: Optical Fibre Embedded into Technical Textile for Healthcare, an Efficient Way to Monitor Patient under Magnetic Resonance Imaging. Proceedings of the 29th Annual International Conference of the IEEE Engineering in Medicine and Biology Society, Lyon, France, 22– 26 August 2007; pp. 3950–3953.

90. Narbonneau, F.; Kinet, D.; Paquet, B.; Depré, A.; de Jonckheere, J.; Logier, R.; Zinke, J.; Witt, J.; Krebber, K. Smart Textile Embedding Optical Fibre Sensors for Healthcare Monitoring during MRI. Adv. Sci. Technol. 2009, 60, 134–143.

91. De jonckheere, J.; Jeanne, M.; Narbonneau, F.; Witt, J.; Paquet, B.; Kinet, D.; Kreber, K.; Logier, R. OFSETH: A Breathing Motions Monitoring System for Patients under MRI. Proceedings of the 32th Annual International Conference of the IEEE Engineering in Medicine and Biology Society, Buenos Aires, Argentina, 31 August–4 September 2010; pp. 1016–1019.

92. De jonckheere, J.; Narbonneau, F.; Jeanne, M.; Kinet, D.; Witt, J.; Krebber, K.; Paquet, B.; Depre, A.; Logier, R. OFSETH: Smart Medical Textile for Continuous Monitoring of Respiratory Motions under Magnetic Resonance Imaging. Proceedings of the 31th Annual International Conference of the IEEE Engineering in Medicine and Biology Society, Minneapolis, USA, 3–6 September 2009; pp. 1473–1476.

93. De Jonckheere, J.; Narbonneau, F.; D'Angelo, L.T.; Witt, J.; Paquet, B.; Kinet, D.; Kreber, K.; Logier, R. FBG-Based Smart Textiles for Continuous Monitoring of Respiratory Movements for Healthcare Applications. Proceedings of the 12th IEEE International Conference on e-Health Networking Applications and Services (Healthcom), Lyon, France, 1–3 July 2010; pp. 277–282.

94. Davis, C.; Mazzolini, A.; Mills, J.; Dargaville, P. A new sensor for monitoring chest wall motion during high-frequency oscillatory ventilation. Med. Eng. Phys. 1999, 21, 619–623.

95. Augousti, A.; Maletras, F.; Mason, J. Improved fibre optic respiratory monitoring using a figure-of-eight coil. Physiol. Meas. 2005, 26, 585–590.

96. Dziuda, L.; Skibniewski, F.; Rozanowski, K.; Krej, M.; Lewandowski, J. Fiber-Optic Sensor for Monitoring Respiration and Cardiac Activity. Proceedings of the 2011 IEEE Sensors Conference, Limerick, UK, 28–31 October 2011; pp. 413–416.

97. Gurkan, D.; Starodubov, D.; Yuan, X. Monitoring of the Heartbeat Sounds Using an Optical Fiber Bragg Grating Sensor. Proceedings of the 2005 IEEE Sensors Conference, Irvine, CA, USA, 30 October–3 November 2005; pp. 306–309.

98. Silva, A.F.; Carmo, J.P.; Mendes, P.M.; Correia, J.H. Simultaneous cardiac and respiratory frequency measurement based on a single fiber Bragg grating sensor. Meas. Sci. Technol. 2011, 22, 075801.

99. Kawamura, M.; Ishizawa, H.; Sato, S.; Koyama, S. Application to Vital Signs by Fiber Bragg Grating Sensing. Proceedings of SICE Annual Conference (SICE), Nagano, Japan, 13–18 September 2011; pp. 2702–2704.

100. Carmo, J.; Ferreira da Silva, A.; Rocha, R.; Correia, H. Application of fiber bragg gratings to wearable garments. IEEE Sens. J. 2012, 12, 261–266.

101. D'Angelo, L.T.; Weber, S.; Honda, Y.; Thiel, T.; Narbonneau, F.; Luth, T.C. A System for Respiratory Motion Detection Using Optical Fibers Embedded into Textiles. Proceedings of the 30th Annual International Conference of the IEEE Engineering in Medicine and Biology Society, Vancouver, BC, Canada, 20–25 August 2008; pp. 3694–3697.

102. Xu, X.B.; Zhang, C.X.; Hou, K.M.; Ding, H.; Diao, X.M. A New Wireless Fiber Optic Sensor for Monitoring Breath Motion. Proceedings of IEEE Conference on New Technologies, Mobility and Security (NTMS), Tangier, Morocco, 5–7 November 2008; pp. 1–5.

103. Pleros, N.; Kanellos, G.T.; Papaioannou, G. Optical Fiber Sensors in Orthopedic Biomechanics and Rehabilitation. Proceedings of the 9th International Conference on Information Technology and Applications in Biomedicine (ITAB), Larnaca, Cyprus, 4–7 November. 2009; pp. 1–4.

104. IASIS Project Description. Available online: http://www.iasis-project.eu (accessed on 12 July 2012).

105. Kanellos, G.T.; Tsiokos, D.; Pleros, N.; Childs, P.; Pissadakis, S. Enhanced Durability FBG-Based Sensor Pads for Biomedical Applications as Human-Machine Interface Surfaces. Proceedings of International Workshop on BioPhotonics, Parma, Italy, 8–10 June 2011; pp. 1–3.

106. Chumanov, E.S.; Remy, C.D.; Thelen, D.G. Tracking the Position of Insole Pressure Sensors during Walking and Running. Proceedings of the 30th Annual Meeting of the American Society of Biomechanics, Palo Alto, CA, USA, 22–25 August 2007.

107. Wertsch, J.J.; Webster, J.G.; Tompkins, W.J. A portable insole plantar pressure measurement system. J. Rehabil. Res. Dev. 1992, 29, 13–18.

108. Cobb, J.; Claremont, D. Transducers for foot pressure measurement: Survey of recent developments. Med. Biol. Eng. Comput. 1995, 33, 525–532.

109. El Kati, R.; Forrester, S.; Fleming, P. Evaluation of pressure insoles during running. Procedia Eng. 2010, 2, 3053–3058.

110. Tappin, J.; Pollard, J.; Beckett, E. Method of measuring 'shearing'forces on the sole of the foot. Clinical Phys. Physiol. Meas. 1980, 1, 83.

111. Lord, M.; Hosein, R.; Williams, R.B. Method for in-shoe shear stress measurement. J. Biomed. Eng. 1992, 14, 181–186.

112. Hao, J.Z.; Tan, K.M.; Tjin, S.C.; Liaw, C.Y.; Chaudhuri, P.R.; Guo, X.; Lu, C. Design of a Foot-Pressure Monitoring Transducer for Diabetic Patients Based on FBG Sensors. Proceedings of the 16th Annual Meeting of the IEEE Lasers and Electro-Optics Society, Tucson, AZ, USA, 27–28 October 2003; pp. 23–24.

113. Wang, W.; Ledoux, W.R.; Sangeorzan, B.J.; Reinhall, P.G. A shear and plantar pressure sensor based on fiber-optic bend loss. J. Rehabil. Res. Dev. 2005, 42, 315–325.

114. Lim, T.K.; Tjin, C.S.; Tay, L.; Chua, C.; Wang, Y.; Brownjohn, J.M.W. Measurement of contact forces between human and vibrating floors using fiber Bragg grating foot sensors. Proc. SPIE 2005, 4235, 161–172.

115. Lebar, A.M.; Harris, G.F.; Wertsch, J.J.; Zhu, H. Development of a Miniature Plantar Shear Force Sensing Transducer. Proceedings of the 15th Annual International Conference of the IEEE Engineering in Medicine and Biology Society, San Diego, CA, USA, 28 – 31 October 1993; pp. 989–990.

116. Akhlaghi, F.; Pepper, M. In-shoe biaxial shear force measurement: The Kent shear system. Med. Biol. Eng. Comput. 1996, 34, 315–317.

117. Koulaxouzidis, A.V.; Holmes, M.J.; Roberts, C.V.; Handerek, V.A. A Shear and Vertical Stress Sensor for Physiological Measurements Using Fibre Bragg Gratings. Proceedings of the 22th Annual International Conference of the IEEE Engineering in Medicine and Biology Society, Chicago, IL, USA, 23–28 July 2000; pp. 55–58.

118. Abu Osman, N.A.; Spence, W.D.; Solomonidis, S.E.; Paul, J.P.; Weir, A.M. The patellar tendon bar! Is it a necessary feature? Med. Eng. Phys. 2010, 32, 760–765.

119. Goh, J.C.H.; Lee, P.V.S.; Chong, S.Y. Comparative study between patellar-tendon-bearing and pressure cast prosthetic sockets. J. Rehabil. Res. Dev. 2004, 41, 491–502.

120. Abu Osman, N.; Spence, W.; Solomonidis, S.; Paul, J.; Weir, A. Transducers for the determination of the pressure and shear stress distribution at the stump–socket interface of trans-tibial amputees. Proc. Inst.Mech. Eng. B J. Eng. Ma. 2010, 224, 1239–1250.

121. Buis, A.; Dumbleton, T.; McHugh, B.; McFadyen, A.; McKay, G.; Murray, K.; Sexton, S. Dynamic interface pressure distributions of two transtibial prosthetic socket concepts. J. Rehabil. Res. Dev. 2009, 46, 405–415.

122. Zhang, M.; Roberts, C. Comparison of computational analysis with clinical measurement of stresses on below-knee residual limb in a prosthetic socket. Med. Eng. Phys. 2000, 22, 607–612.

123. Portnoy, S.; Yarnitzky, G.; Yizhar, Z.; Kristal, A.; Oppenheim, U.; Siev-Ner, I.; Gefen, A. Real-time patient-specific finite element analysis of internal stresses in the soft tissues of a residual limb: A new tool for prosthetic fitting. Ann. Biomed. Eng. 2007, 35, 120–135.

124. Moo, E.K.; Osman, N.A.; Pingguan-Murphy, B.; Abas, W.; Spence, W.D.; Solomonidis, S.E. Interface pressure profile analysis for patellar tendon-bearing socket and hydrostatic socket. Acta Bioeng. Biomed. 2009, 11, 37–43.

125. Dou, P.; Jia, X.; Suo, S.; Wang, R.; Zhang, M. Pressure distribution at the stump/socket interface in transtibial amputees during walking on stairs, slope and non-flat road. Clin. Biomech. 2006, 21, 1067–1073.

126. Kanellos, G.T.; Papaioannou, G.; Tsiokos, D.; Mitrogiannis, C.; Nianios, G.; Pleros, N. Two dimensional polymer-embedded quasi-distributed FBG pressure sensor for biomedical applications. Opt. Express 2010, 18, 179–186.

127. Papaioannou, G.; Mitrogiannis, C.; Nianios, G.; Fiedler, G. Towards a Novel "SMART Socket" Design for Lower Extremity Amputees. Proceedings of the HFM Symposium on Use of Advanced Technologies and New Procedures in Medical Field Operations, Essen, Germany, 19–21 April 2010.

128. Papaioannou, G.; Mitrogiannis, C.; Fiedler, G.; Nianios, G. Assessment of Vacuum-Assisted Trans-Tibial Amputee Socket Dynamics. Proceedings of the 9th International Conference on Information Technology and Applications in Biomedicine, Lamaca, Cyprus, 5–7 November 2009; pp. 1–4.

129. Center for Applied Biomechanics. University of Malaya. Available online: cab.um.edu.my (accessed on 12 July 2012).

130. Photonics Research Group. University of Malaya. Available online: photonics.um.edu.my (accessed on 28 August 2012).

131. Hao, J.; Jayachandran, M.; Kng, P.L.; Foo, S.F.; Aung Aung, P.W.; Cai, Z. FBG-based smart bed system for healthcare applications. Front. Optoelectron. China 2010, 3, 78–83.

132. Fleming, B.C.; Beynnon, B.D. In vivo measurement of ligament/tendon strains and forces: A review. Ann. Biomed. Eng. 2004, 32, 318–328.

133. Vilimek, M. Using a fiber Bragg grating sensor for tendon force measurements. J. Biomech. 2008, 41, S511.

134. Ren, L.; Song, G.; Conditt, M.; Noble, P.C.; Li, H. Fiber Bragg grating displacement sensor for movement measurement of tendons and ligaments. Appl. Opt. 2007, 46, 6867–6871.

135. De Rossi, D.; Lorussi, F.; Mazzoldi, A.; Orsini, P.; Scilingo, E.P. Monitoring Body Kinematics and Gesture through Sensing Fabrics, Proceedings of the 1st Annual International Conference On Microtechnologies in Medicine and Biology, Lyon, France, 12–14 October 2000; pp. 587–592.

136. Lorussi, F.; Rocchia, W.; Scilingo, E.P.; Tognetti, A.; De Rossi, D. Wearable, redundant fabric-based sensor arrays for reconstruction of body segment posture. IEEE Sens. J. 2004, 4, 807–818.

137. Rocha, L.A.; Correia, J.H. Wearable Sensor Network for Body Kinematics Monitoring. Proceedings of the 10th IEEE International Symposium on Wearable Computers, Montreux, Switzerland, 11–14 October 2006; pp. 137–138.

138. Yamamoto, Y. An alternative approach to the acquisition of a complex motor skill: Multiple movement training on tennis strokes. Int. J. Sport Health Sci. 2004, 2, 169–179.

139. Shany, T.; Redmond, S.J.; Narayanan, M.R.; Lovell, N.H. Sensors-based wearable systems for monitoring of human movement and falls. IEEE Sens. J. 2012, 12, 658–670.

140. Werner, M.R.; Fahrner, W.R. Review on materials, microsensors, systems and devices for high-temperature and harsh-environment applications. IEEE Trans. Ind. Electron. 2001, 48, 249–257.

141. da Silva, A.F.; Goncalves, A.F.; de Almeida Ferreira, L.A.; Araujo, F.M.M.; Mendes, P.M.; Correia, J.H. PVC smart sensing foil for advanced strain measurements. IEEE Sens. J. 2010, 10, 1149–1155.

142. Ferreira da Silva, A.; Goncalves, A.F.; de Almeida Ferreira, L.A.; Araujo, F.M.M.; Mendes, P.M.; Correia, J.H. A smart skin PVC foil based on FBG sensors for monitoring strain and temperature. IEEE Trans. Ind. Electron. 2011, 58, 2728–2735.

143. Ferreira da Silva, A.; Goncalves, F.; Mateus Mendes, P.; Higino Correia, J. PVC formulation study for the manufacturing of a skin smart structure based in optical fiber elements. Polym. Advan. Technol. 2012, 23, 220–227.

144. Silva, A.F.; Mendes, P.M.; Correia, J.H.; Goncalves, F.; Ferreira, L.A.; Araujo, F.M. Smart Sensing Polymeric Foil with Integrated Optic Fiber Sensors: Fabrication and Characterization of a Polymeric Foil Sensitive to Strain. Proceedings of the 3rd International Conference on Sensor Technologies and Applications, Glyfada, Athens, Greece, 18–23 June 2009; pp. 90–93.

145. Silva, A.F.; Goncalves, F.; Ferreira, L.A.; Araujo, F.M.; Dias, N.S.; Carmo, J.P.; Mendes, P.M.; Correia, J.H. Manufacturing Technology for Flexible Optical Sensing Foils. Proceedings of the 35th Annual Conference of IEEE Industrial Electronics (IECON), Porto, Portugal, 3–5 November 2009; pp. 1883–1886.

146. Silva, A.; Gonçalves, F.; Ferreira, L.A.A.; Araújo, F.M.; Mendes, P.; Correia, J.H. Fiber Bragg grating sensors integrated in polymeric foils. Mater. Sci. Forum 2010, 636, 1548–1554.

147. Rocha, R.P.; Silva, A.F.; Carmo, J.P.; Correia, J.H. FBG in PVC Foils for Monitoring the Knee Joint Movement during the Rehabilitation Process. Proceedings of the 33th Annual International Conference of the IEEE Engineering in Medicine and Biology Society, Boston, MA, USA, 30 August–3 September 2011; pp. 458–461.

148. Nishiyama, M.; Watanabe, K. Wearable sensing glove with embedded hetero-core fiber-optic nerves for unconstrained hand motion capture. IEEE Trans. Instrum. Meas. 2009, 58, 3995–4000.

149. Chin-Shyurng, F.; Herman, S. Development of a data glove with reducing sensors based on magnetic induction. IEEE Trans. Ind. Electron. 2005, 52, 585–594.

150. Perng, J.K.; Fisher, B.; Hollar, S.; Pister, K.S.J. Acceleration Sensing Glove (ASG). Proceedings of the 3rd International Symposium on Wearable Computers, San Francisco, CA, USA, 18–19 October 1999; pp. 178–180.

151. Enzo Pasquale, S.; Lorussi, F.; Mazzoldi, A.; De Rossi, D. Strain-sensing fabrics for wearable kinaesthetic-like systems. IEEE Sens. J. 2003, 3, 460–467.

152. Lorussi, F.; Enzo Pasquale, S.; Tesconi, M.; Tognetti, A.; De Rossi, D. Strain sensing fabric for hand posture and gesture monitoring. IEEE Trans. Inf. Technol. Biomed. 2005, 9, 372–381.

153. da Silva, A.F.; Goncalves, A.F.; Mendes, P.M.; Correia, J.H. FBG sensing glove for monitoring hand posture. IEEE Sens. J. 2011, 11, 2442–2448.

CHAPTER 7

Optical Fibre Long-Period Gratings Functionalised with Nano-Assembled Thin Films: Approaches to Chemical Sensing

Sergiy Korposh[1], Seung-Woo Lee[1], Stephen James[2] and Ralph Tatam[2]

[1] Graduate School of Environmental Engineering, the University of Kitakyushu, 1-1 Hibikino, Wakamatsu-ku, Kitakyushu, Japan

[2] School of Engineering, Cranfield University, Cranfield, Bedford, UK

1. INTRODUCTION

Optical techniques are considered as powerful tools for the development of chemical and biological sensors, covering a wide range of applications including bio-chemical and food analysis, and environmental and industrial monitoring [1, 2, 3]. Optical fibre sensors, as a result of their advantages such as high sensitivity, compactness, remote measurement and multiplexing capabilities, have attracted a great deal of attention for the development of refractometers and chemical sensors [4, 5, 6]. Refractometers and chemical sensors based on optical fibre gratings, both fibre Bragg gratings (FBGs) and long period gratings (LPGs), have been extensively employed for refractive index measurements and monitoring associate chemical processes since they offer wavelength-encoded information, which overcomes the referencing issues associated with intensity based approaches.

FBG based approaches have exploited side polished [7] and thinned [8,9] optical fibres to expose the evanescent field of the mode propagating in the core of the fibre to the surrounding medium, such that the Bragg wavelength becomes sensitive to the surrounding refractive index. The response of such sensors is non-linear, with a maximum predicted sensitivity of approximately 200 nm/refractive index unit (RIU) for indices close to that of the cladding of the optical fibre [8]. Experimental investigation revealed a sensitivity of 70 nm/RIU with a corresponding resolution of order 10^{-4} RIU, assuming that the Bragg wavelength is measured with a resolution of 1 pm [9].

The effective refractive indices of the modes of the cladding of the optical fibre are inherently sensitive to the surrounding refractive index. Tilted fibre Bragg gratings (TFBGs) and LPGs allow the controlled, resonant coupling of light from the core of the optical fibre into cladding modes. The characteristics of the resonant coupling are modified by changes in the surrounding refractive index, affording the ability to form refractive index sensors without altering the geometry of the optical fibre.

TFBGs facilitate the coupling of the propagating core mode to backward propagating cladding modes. As the coupling wavelength and efficiencies are dependent upon the properties of the cladding modes, the resonance features in the TFBG transmission spectrum exhibit sensitivity to surrounding refractive index [10,11]. Analysis of the transmission spectrum facilitates measurements with resolution of 10^{-4} RIU [11]. The use of thin film coatings of refractive index higher than that of the cladding on both polished FBGs and TFBGS has been shown to allow the RI range over which the devices show their highest sensitivity to the tuned to lower values [10,12,13].

LPGs promote coupling between the propagating core mode and co-propagating cladding modes, i.e. work as transmission gratings and are more attractive for practical applications as compared to FBGs, owing to lower cost of analytical instruments used to interrogate them. The high attenuation of the cladding modes results in the transmission spectrum of the fibre containing a series of attenuation bands centred at discrete wavelengths, each attenuation band corresponding to the coupling to a different cladding mode, as shown in Figure 1 [14].

The refractive index sensitivity of LPGs arises from the dependence of the phase matching condition upon the effective refractive index of the cladding modes that is governed by Equation 1 [14]:

$$\lambda_{(x)} = (n_{core} - n_{clad(x)})\Lambda \tag{1}$$

where $\lambda_{(x)}$ represents the wavelength at which the coupling occurs to the linear polarized (LP_{0x}) mode, n_{core} is the effective RI of the mode propagating in the core, $n_{clad(x)}$ is the effective RI of the LP_{0x} cladding mode, and Λ is the period of the grating.

The effective indices of the cladding modes are dependent upon the difference between the refractive index of the cladding and that of the medium surrounding the cladding. The central wavelengths of the attenuation bands thus show a dependence upon the refractive index of the medium surrounding the cladding, with the highest sensitivity being shown for surrounding refractive indices close to that of the cladding of the optical fibre, provided that the cladding has the higher refractive index [15]. For surrounding refractive indices higher than that of the cladding, the centre wavelengths of the resonance bands show a considerably reduced sensitivity [15].

The refractive index sensitivity of an LPG is dependent upon the order of the cladding mode that is coupled to, allowing the tuning of the sensitivity by appropriate choice of grating period, with 427.72, 203.18, 53.45, and 32.10 nm/RIU being reported for LPGs fabricated in single mode fibre (SMF) 28 with period 159, 238, 400, and 556 μm, respectively [16]. A further consideration is the geometry and composition of the fibre, with the sensitivity being shown to differ for step index and W profile fibres and a progressive three layered fibre [17]. New fibre geometries, such as photonic crystal fibres [18,19] and photonic band gap fibres [20] have also been investigated for the measurement of the refractive index of a liquid that fills the air channels. LPGs in liquid filled solid-core photonic bandgap fibre have been shown to exhibit a sensitivity of 17,900 nm/RIU to changes in the RI of the liquid [21], but the requirement to fill the fibre with the liquid of interest may limit application as a refractometer.

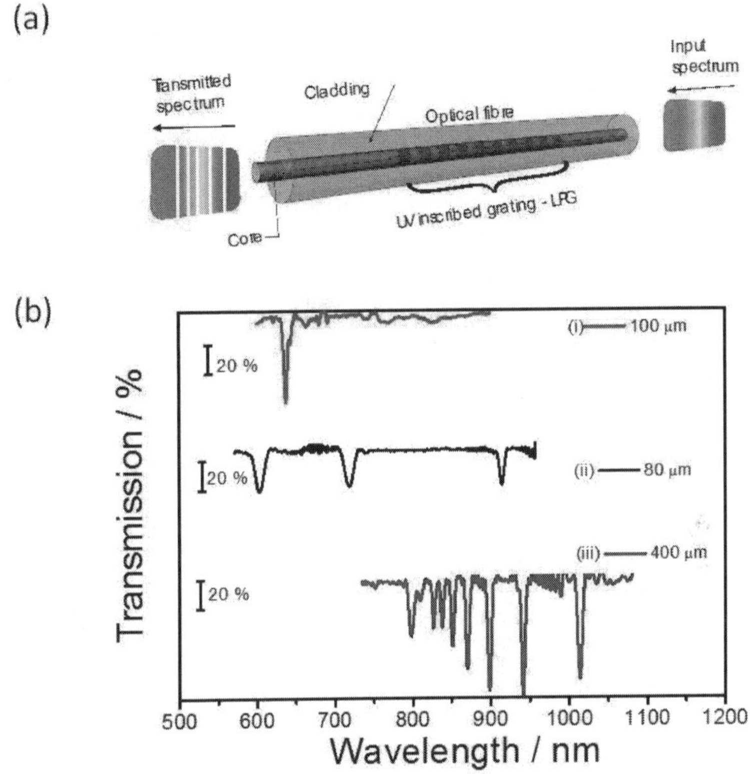

Figure 1. a) Schematic illustration of the LPG structure and (b) transmission spectra of LPGs with different grating periods fabricated in an optical fibre of cut-off wavelength 670 nm (Fibrecore sm700): (i) 80 µm, (ii) 100 µm, and (iii) 400 µm [5].

In order to improve the sensitivity of the LPGs written in standard optical fibre configurations to surrounding RI, approaches such as tapering the fibre [22] or etching the cladding [23,24] have been investigated. Tapering the fibre to a diameter of 25 µm allowed the demonstration of an LPG with a sensitivity of 715 nm/RIU [22]. Etching the cladding of a section of standard single mode fibre containing an LPG from 125 µm to approximately 100 µm produced a sensitivity gain of 5 [25], while the etching an arc induced LPG to a diameter of 37 µm allowed the demonstration of a sensitivity of order 20,000 nm/RIU [26]. Approaches that require the processing of the fibre, such as polishing, etching and tapering, produce significant enhancements in sensitivity, but at the cost of requiring careful packaging to compensate for the inevitable reduction in the mechanical integrity of the device.

The deposition of thin film overlays, of thickness on the order of 200 nm, of materials of refractive index higher than that of the cladding has also been investigated for the enhancement of refractive index sensitivity [25,26]. It has been shown previously experimentally and theoretically [27,28] that the effective indices of the cladding modes, and thus the central wavelengths of the core-cladding mode coupling bands of LPGs, show a high sensitivity to the optical thickness of high refractive index coatings when the coating's optical thickness is such that one of the low order cladding modes is phase matched to a mode of the waveguide formed by the coating. This is termed the *mode transition region*, in which the cladding modes are reorganized, with each mode taking on the characteristics (effective index and electric field profile) of its adjacent lower order cladding mode [29]. The output from a numerical model of the influence of coating thickness on the effective indices of the cladding modes of an optical fibre is plotted in Figure 2, showing clearly the mode transition region.

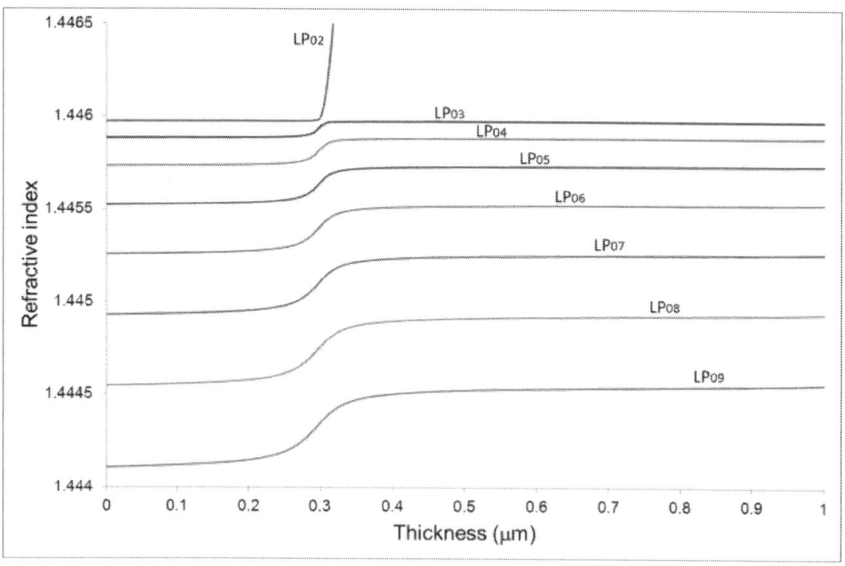

Figure 2. Plot of the effective refractive indices of the first 9 cladding modes as a function of the thickness of an overlay of refractive index. The model assumed values for the core and cladding refractive indices of 1.4496 and 1.446 respectively, a core radius of 3.5 μm, and that the refractive index of the overlay material was 1.57.

Biasing the optical thickness of the coating such that the LPG is operating in the *mode transition region* enhances the sensitivity of the cladding mode effective indices and thus the resonance wavelengths to the surrounding refractive index. A theoretical analysis explored the optimization of the refractive index sensitivity by selecting grating period, coating thickness and refractive index, predicting a sensitivity of 5980 nm/RIU [29].

In addition, the combination of optical fibers and nanomaterials provides a prospect for the fabrication of chemical sensors with high sensitivity and that offer specific response to targeted chemical species [30,31]. Achieving the coating thickness that provides optimized sensitivity requires control on the nm scale, which is why many reports have exploited the Langmuir Blodgett (LB) and layer-by-layer (LbL) electrostatic self assembly (ESA) coating deposition techniques, where a multilayer coating is deposited with each layer having a thickness of order 1 nm. Based on this principle, sensors for organic vapors, metal ions, humidity, organic solvents and biological materials have been reported [32, 33, 34]. Similarly to surface plasmon resonance (SPR) devices, LPG sensors can provide highly precise analytical information about adsorption and desorption processes associated with the RI and thickness of the sensing layer. For instance, the sensitivity of LPG sensors is in the same order of magnitude as SPR sensors, showing a sensitivity of ca. 1 nM for antigen–antibody reactions. An advantage of the LPG over SPR lies in the ability to fabricate a cheap and portable device that can be applied in various analytical situations.

The influence of the period of the LPG on the sensitivity of the mode coupling to perturbations such as changes in surrounding RI can be understood with reference to the phase matching condition, equation (1). Using the weakly guided approximation it is possible to determine the effective indices of the modes of the core and cladding, and, using equation 1, to generate a family of phase matching curves that describe the variation of the resonant wavelength with period of the LPG, an example is shown in Figure 3. It can be seen that the phase matching condition for each cladding mode contains a turning point. The resonance features in the LPG spectrum exhibit their highest sensitivity to external perturbations near the phase matching turning point. The response of the transmission spectrum of an LPG operating near the phase matching turning point external perturbation, in this case changes to the optical thickness of a coating deposited onto the optical fibre, are illustrated in Figure 4. In Figure 4a, the grating period is such that phase matching to the LP_{020} cladding mode is satisfied, but it is not possible to couple to the LP_{021} cladding mode, with a resulting LPG transmission spectrum of the form shown in Figure 4b. Changes in the optical thickness (product of the geometrical thickness and refractive index) of the coating cause an increase in the effective refractive index of the cladding modes, and the phase matching curves change accordingly, as illustrated in Figure 4c, resulting in the development of a resonance band corresponding to coupling to the LP_{021} cladding mode, and a small blue shift in the central wavelength of the LP_{020} resonance band, Figure 4d. Further increases in the optical thickness of the coating result in the further development of the LP_{021} resonance band, which subsequently splits into two bands, the so called dual resonance, Figures 4f and 4h. The small gradient of the phase matching curve at the phase matching turning point results in the LP_{021} resonance band exhibiting much higher sensitivity than the LP_{020} resonance band for this grating period. Thus the sensitivity of coated LPG sensors can be optimised by appropriate choice of the optical thickness of the coating and period of the grating to ensure that the LPG operates at both the mode transition region and the phase matching turning point.

Recently, LPG fibre sensors with porous coatings have attracted a lot of interest. For instance, a sol-gel film of SnO_2 of thickness of order 200 nm was deposited onto to an LPG, facilitating the demonstration of an ethanol gas sensor. The porosity and high RI of the coating material resulted in the LPG spectrum exhibiting a response to the diffusion of ethanol gas into the coating and the authors predicted that an optimized sensor would exhibit a detection limit of 100 ppb. Sol-gel coatings of SiO_2 and TiO_2 have been deposited onto LPGs, revealing a gain of 2 in the sensitivity to external RI, with the higher index TiO_2 coating offering the larger response (up to 1067.15 nm/RIU). The authors noted that the higher the index and thickness of the coating the more pronounced was the enhancement in LPG sensitivity compared to the equivalent uncoated LPG [16].

Recently we have demonstrated a fibre optic refractive index sensor based on a long period grating (LPG) with a nano-assembled mesoporous coating of alternate layers of poly(allylamine hydrochloride) (PAH) and SiO_2 nanospheres [5, 6]. PAH/SiO_2 coatings of different thicknesses were deposited onto an LPG operating near its phase matching turning point in order to study the effect of the film thickness and porosity on sensor performance. The device showed a high sensitivity (1927 nm/RIU) to RI changes with a response time less than 2 sec over a wide RI range (1.3330–1.4906). The low refractive index of the mesoporous film, 1.20@633 nm, facilitates the measurement of external indices higher than that of the cladding, extending the range of operation of LPG based RI sensors. The ability of this device to monitor, in real time, RI changes during a dilution process was also demonstrated.

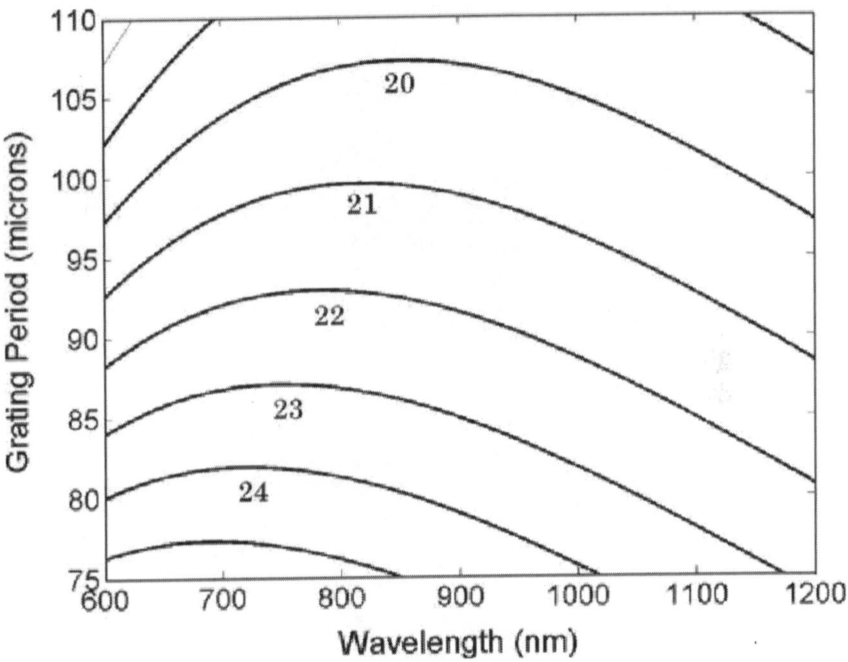

Figure 3. Phase matching curves for higher order cladding modes of an optical fibre of cut off wavelength 670 nm. The numbers refer to the order of the cladding mode.

In this chapter we introduce a new approach to LPG based chemical sensing. A novel 2 stage approach to the development of the sensor is explored. The first stage involves the deposition of the mesoporous coating onto the LPG operating near the phase matching turning point. In the second stage a functional material, chosen to be sensitive to the analyte of interest, is infused into the base mesoporous coating. The mesoporous coating consists of a multilayer film of SiO_2 nanoparticles (SiO_2 NPs) deposited using the LbL technique. The initially low RI of the mesoporous coating, 1.2@633 nm, is increased significantly by the chemical infusion, resulting in a large change in the LPG's transmission spectrum. The sensing of ammonia in aqueous solution was chosen to demonstrate the sensing principle. The operation of the sensor was characterised using two functional materials, tetrakis-(4-sulfophenyl)porphine (TSPP) and polyacrylic acid (PAA). TSPP is a porphine compound that changes its optical properties (absorbance and RI) in response to exposure to ammonia, while PAA has been employed as a functional compound ammonia binding [35]. In the case of the PAA it is assumed that direct biding of ammonia to the COOH moiety will change RI of the mesoporous coating, while in the case of TSPP its desorption will result on RI change.

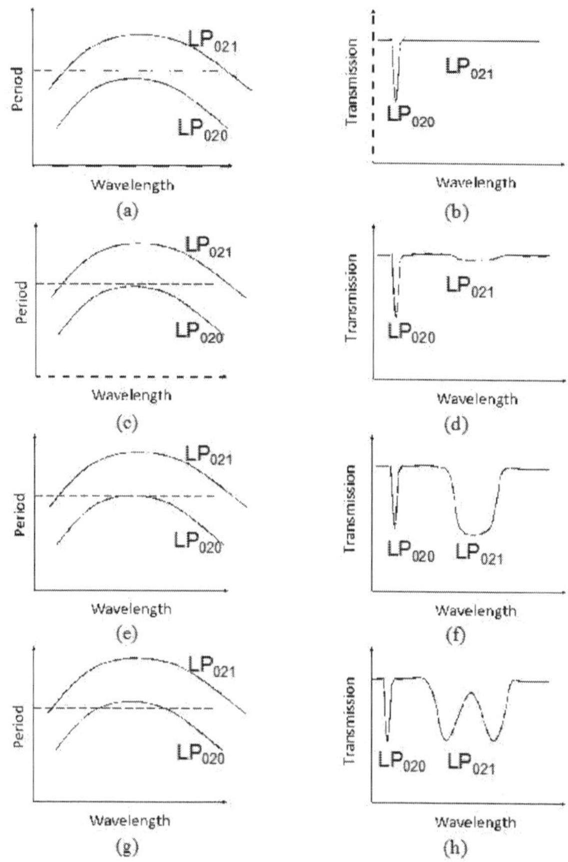

Figure 4. Optical features of the LP_{020} and LP_{021} cladding modes near the phase matching turning point. An increase in the effective index of the cladding modes, caused for example by an increase in the surrounding refractive index, or in the optical thickness of a coating deposited onto the fibre, cause the phase matching curves to move as shown, producing large changes in the coupling characteristics and transmission spectrum.

2. SENSOR FABRICATION

Tetrakis-(4-sulfophenyl)porphine (TSPP, Mw = 934.99) was purchased from Tokyo Kasei, Japan. Poly(diallyldimethylammonium chloride) (PDDA, Mr = 200,000–350,000, 20% (w/w) in H2O; monomer Mw = 161.5 g mol−1), PAA_{25} (Mw: 250,000, 35 wt% in H_2O) and ammonia 30 wt% aqueous solution were purchased from Aldrich. A colloidal solution of silica nanoparticles (SiO_2 NPs), SNOWTEX 20L (40–50 nm), was purchased from Nissan Chemical. All chemicals were reagents of analytical grade, and used without further purification. Deionized pure water (18.3 MΩ cm) was obtained by reverse osmosis followed by ion exchange and filtration in a Millipore-Q (Millipore, Direct-QTM).

A detailed description and reference to the optical properties of LPGs can be found elsewhere [5, 6, 36]. In this work, an LPG of length 30 mm with a period of 100 μm was fabricated in a single mode optical fibre (Fibercore SM750) with a cut-off wavelength of 670 nm using point-by-point UV writing process. The photosensitivity of the fibre was enhanced by pressurizing it in hydrogen for a period of 2 weeks at 150 bar at room temperature.

SiO_2 NPs were deposited onto the surface of the fibre using the LbL process, as illustrated in Figure 5a. As the LPG transmission spectrum is known to be sensitive to bending, for the film deposition process and ammonia detection experiments the optical fibre containing LPG was fixed within a special holder, as shown in Figure 5b, such that the section of the fibre containing the LPG was taut and straight throughout the experiments [36]. The detailed procedure of the deposition of the SiO_2 NPs onto the LPG and infusion of the TSPP compound has been previously reported [5]. Briefly, the section of the optical fibre containing LPG, with its surface treated such that it was terminated with OH groups, was alternately immersed into a 0.5 wt% solution containing a positively charged polymer, PDDA, and, after washing, into a 1 wt% solution containing the negatively charged SiO_2 NPs solution, each for 20 min. This process was repeated until the required coating thickness was achieved. When the required film thickness had been achieved (i.e. when the development of the second resonance band was observed with the fibre immersed into water), ca. after 10 deposition cycles, the coated fibre was immersed in a solution of TSPP or PAA as functional compound for 2 h, which was infused into the porous coating and provided the sensor with its specificity. Due to the electronegative sulfonic groups present in the TSPP compound, an electrostatic interaction occurs between

TSPP and positively charged PDDA in the PDDA/SiO_2 film. On the other hand, PAA is usually considered as a promising sensor element for ammonia sensing, owing to the presence of free carboxylic function groups that lead to high affinity towards amine compounds [37]. After immersion into the TSPP and PAA solutions, the fibre was rinsed in distilled water, in order to remove physically adsorbed compounds, and dried by flushing with N_2 gas. The compounds remaining in the porous silica structure were bound to the surface of the polymer layer that coated each nanosphere. This effectively increased the available surface area for the compounds to bond to. The presence of functional chemical compounds increased the RI of the porous coating and resulted in a significant change in the LPG's transmission spectrum, consistent with previous observations for increasing the coating thickness [38]. All experiments have been conducted at 25°C and 50% of rH.

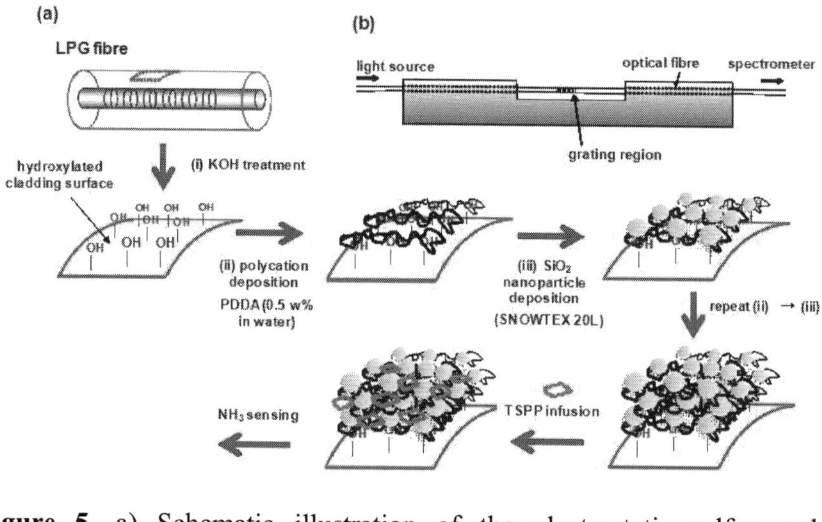

Figure 5. a) Schematic illustration of the electrostatic self-assembly deposition process and (b) deposition cell with a fixed LPG fibre.

3. MESOPOROUS COATING OF LONG PERIOD GRATINGS

Figures 6a and 6b show the surface morphology and cross-section of the 10-cycle (PDDA/SiO_2) layer, referred to as $(PDDA/SiO_2)_{10}$, on a quartz substrate, respectively. As can be seen, the $(PDDA/SiO_2)_{10}$ film has a uniform surface consisting of SiO_2 NPs with an average diameter of 45 nm (Figure 6a). The film thicknesses obtained after 1 and 10 cycles of the ($PDDA/SiO_2$) deposition process, determined from SEM cross-section measurements, are approximately 50±2 nm and 450±20 nm, respectively (Figures 6b and 6c). The pore size distribution of the ($PDDA/SiO_2$) film indicates a well-developed mesoporous structure with a mean pore radius of 12.5 nm and specific surface area of 50 m^2 g^{-1} [5].

The transmission spectrum of the 100 μm period LPG undergoes changes due to the alternate deposition of SiO_2 NPs, as shown in Figure 7, which influences the effective RI of the cladding mode, as described previously [5]. When the LPG was in the silica colloidal solution, the resonance feature (at ca. 640 nm) corresponding to coupling to the LP_{020} cladding mode exhibited a small blue wavelength shift of 8.5 nm. As the optical thickness of the coating increased, it became possible to couple energy to the LP_{021} mode, with the corresponding development of the resonance band at ca. 800 nm [5], at the phase matching turning point for that mode. However, because of the low RI (1.20) [39] of the porous silica coating, this resonance feature is not well developed in water and is not present in air for this coating thickness.

When an uncoated LPG fibre with a period of 100 μm is immersed into water (RI=1.323), a blue shift of the LP_{020} resonance band of 3 nm is observed. However, the sensitivity of the LP_{020} resonance band is much improved by coating the LPG with a $(PDDA/SiO_2)_{10}$ film, showing a blue shift of 7 nm when the LPG was immersed in water, along with the appearance of the well pronounced second resonance band, LP_{021} (Figure 8a). This is attributed to the increase in sensitivity of the LPG to the surrounding RI [16], being of interest when measuring the RI of low concentration aqueous solutions. The response to the RI changes is fast (< 2 s) and stable, as indicated by the measurement of the transmitted power at the centre of the LP_{021} resonance (Figure 8b).

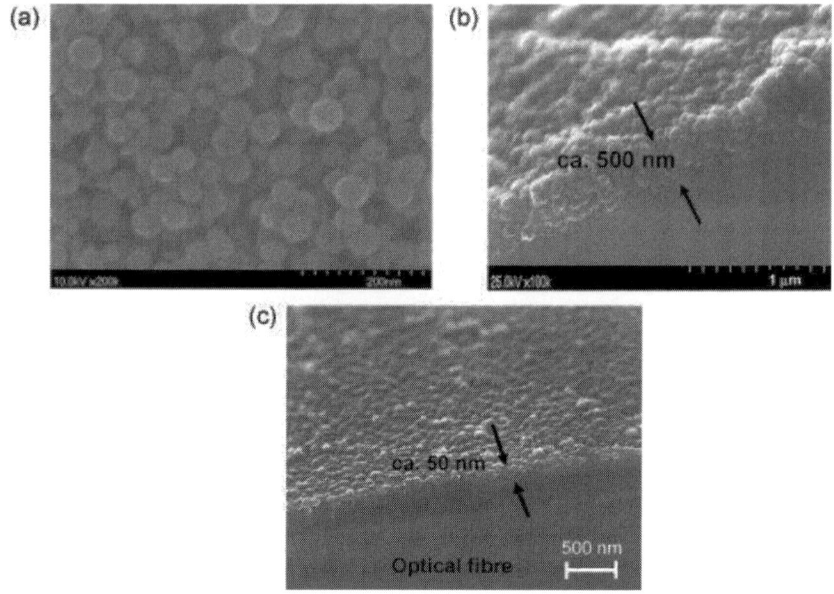

Figure 6. SEM images of the (a) surface morphology and (b) cross-section of the $(PDDA/SiO_2)_{10}$ film deposited on a quartz substrate before TSPP infusion; (c) cross section of a $(PDDA/SiO_2)_1$ film.

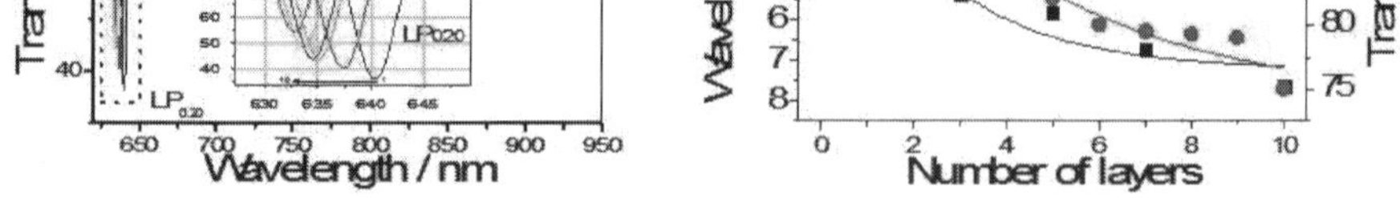

Figure 7. a) Transmission spectra of the 100 μm period LPG in the colloidal SiO_2 solution in water after PDDA deposition (b) wavelength shifts and changes in transmission as a function of the number of deposition cycles for the LP_{020} and LP_{021} resonance bands, respectively; the curve is a guide to the eye only.

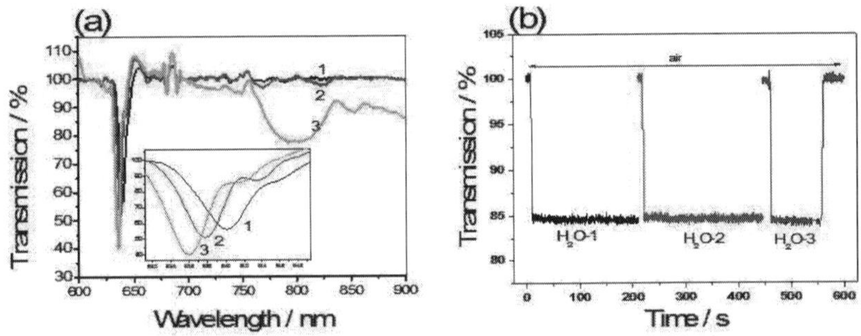

Figure 8. a) Transmission spectra of the 100 μm period LPG under different conditions: black line, in air without coating; red line, in water without coating; green line, in water after deposition of the $(PDDA/SiO_2)_{10}$ film. (b) Dynamic changes of the transmission spectrum of the SiO_2 NP coated LPG measured at 800 nm in different phases from air to water.

The RI and film thickness of a 1-cycle $PDDA/SiO_2$ film, measured using ellipsometry, were 1.20 (at 633 nm) and 47±2 nm, respectively, which is in a good agreement with both the reported data for RI [39] and the thickness measured using SEM (see Figure 6b). It should be noted that dispersion of the RI in the wavelength range of 400–800 nm is negligible, with the RI value of 1.20±0.0001, and does not influence LPG sensor performance over the operational wavelength range of 600–900 nm. The deposition of the $PDDA/SiO_2$ layer was also monitored using the QCM technique and UV-vis spectroscopy (data not shown). A $(PDDA/SiO_2)_{10}$ film was deposited on two different QCM electrodes and a relative standard deviation of ±11% was obtained. The frequency linearly decreased as the number of the deposition cycles increased with average values of 1739±207 Hz and 30±10 Hz for the SiO_2 and PDDA layers, respectively. This corresponds to SiO_2 and PDDA masses of 1565 ng and 27 ng per each layer, respectively. The average thickness of the SiO_2 layer was 42±4 nm [36], which corresponds very well with the values determined using ellipsometry and SEM. A uniform $PDDA/SiO_2$ film was assembled on the quartz substrate, as revealed by the increase of the absorbance in the UV region with increment in the number of SiO_2 NP layers. The modulation observed in the absorption spectrum of the coated quartz slide was the result of the interference between light reflected from the front and rear surfaces of the coated slide, which introduces a channelled spectrum, the period and phase of which is dependent on the $PDDA/SiO_2$ film optical thickness. Thickness values of the $PDDA/SiO_2$ film deposited onto different substrates (quartz slab, silicon wafer and optical fibre) using a LbL method determined from SEM, QCM and ellipsometry measurements agreed well, within experimental error, regardless of the surface nature, indicating that the LbL method is an efficient tool for the deposition of uniform nano-thin films on different types of surfaces.

4. EMPLOYMENT OF FUNCTIONAL MOLECULES AND CHEMICAL SENSING

The principle of operation of a coated LPG sensor operating at the phase matching turning point and applied to the detection of chemical components that can be bound through an electrostatic interaction with PDDA in the mesopores of the film has been discussed previously [5]. The resonance bands (LP_{020} and LP_{021}) could be used for the detection of chemical components that can be bound through an electrostatic interaction with the cationic groups of PDDA in the mesopores of the film. When an LPG coated with a $(PDDA/SiO_2)_{10}$ film is immersed in a TSPP solution, the transmission spectrum undergoes significant changes. Figure 9a shows the transmission spectra recorded when the $(PDDA/SiO_2)_{10}$ film coated LPG was immersed in a 1 mM TSPP water solution. As the TSPP is infused into the film, the RI of the film increases (from 1.200 to ca. 1.540, measured using ellipsometry) and the phase matching condition for coupling to LP_{021} is satisfied. A broad single attenuation band is developed rapidly (within 60 s), which subsequently splits in two bands as the RI of the coating increases in response to the TSPP infusion. The required time to complete the binding between the TSPP and PDDA moieties is less than 600 s (Figure 9b). The observed response indicates a large increase in the optical thickness of the film, which is a result of the increase in the RI of the film, as the TSPP is infused into the porous structure and adsorbed to the PDDA moiety between SiO_2 NPs.

The evolution of the transmission spectrum of the SiO_2 coated LPG when immersed in the TSPP solution is shown in the grey scale plot shown in Figure 10, where the transmission is represented by white and black, corresponding to 100% and 0%, respectively. The dark line at around 635 nm, which originates at a wavelength of 640 nm in the uncoated LPG, represents the resonance band that corresponds to the first order coupling to the LP_{020} cladding mode. The discontinuity in the trace, at 60 s, occurred when the LPG was immersed in the solution and is a result of the LPG's sensitivity to the RI of the solution.

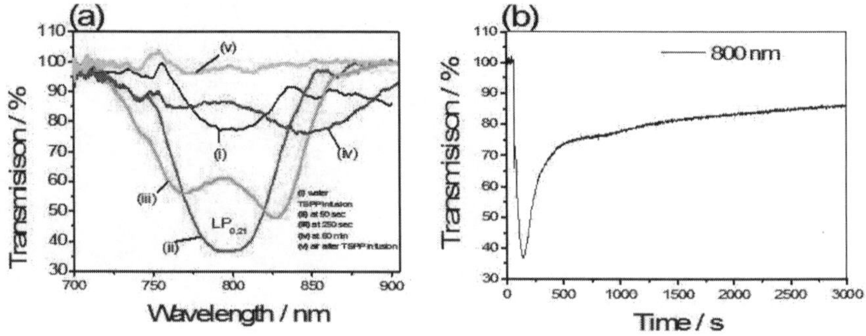

Figure 9. a) Transmission spectra of the SiO_2 NP coated LPG and (b) the dynamic transmission change recorded at 800 nm when the SiO_2 NP coated LPG was immersed in a TSPP solution (1 mM in water).

In order to assess the sensitivity of the optical device, the $(PDDA/SiO_2)_{10}$ coated LPG was exposed to different concentrations of TSPP, and the results are shown in Figures 11a and 11b. The increase of the TSPP concentration from 10 to 1000 μM results in a decrease of the transmission measured at 800 nm, corresponding to the development of the LP_{021} cladding mode resonance. This is also accompanied by a blue shift of the LP_{020} resonance band, indicating the increase of the RI of the film. The response time of the sensor is observed to be slower at lower TSPP concentrations. For 1 mM TSPP concentration, the increase in transmission at 800 nm, shown in Figures 11b, is attributed to the splitting of the fully developed LP_{021} cladding mode resonance into dual resonance bands.

The ability to reuse the device was tested by removing the infused TSPP molecules from the film using an ammonia solution (ca. 1000 ppm). The spectrum was reverted to that observed for the $(PDDA/SiO_2)_{10}$ coated LPG before TSPP infusion. Subsequent immersion of the $(PDDA/SiO_2)_{10}$ coated LPG into the TSPP solutions allowed the results shown in Figure 11 to be reproduced.

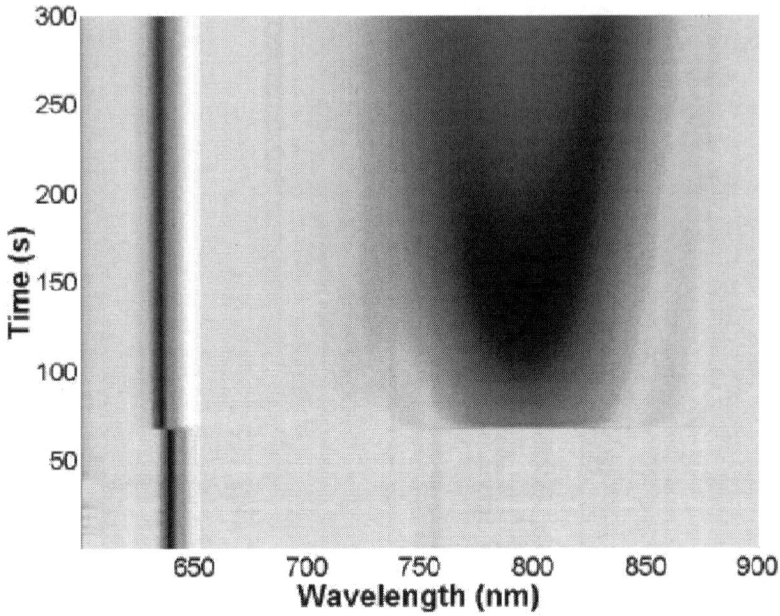

Figure 10. The evolution of the transmission spectrum of the SiO_2 NP coated LPG (period 100 μm), when immersed in an aqueous solution of TSPP (1 mM). The grey scale represents the measured transmission, with white corresponding to 100%, and black to 0%.

A similar effect to that observed for the infusion of TSPP was observed when the $(PDDA/SiO_2)_{10}$ coated LPG was immersed into a PAA solution. The magnitude of the change, however, was smaller as compared to that induced by TSPP infusion, most plausibly being related to the lower RI of PAA (1.442) [40] as compared with that of TSPP (1.540) [5]. It should be noted that thickness of the $(PDDA/SiO_2)_{10}$ film was not changed on the infusion of the functional compounds, as revealed from SEM cross-sectional and ellipsometry measurements of the samples deposited onto the optical fibre before and after TSPP infusion (data not shown).

The infusion of the TSPP molecules into the $PDDA/SiO_2$ film deposited on a glass substrate was also investigated using UV-vis spectroscopy. Figure 12 shows the absorbance spectrum of the $PDDA/SiO_2$ film after infusion of the TSPP compound. As can be seen from Figure 12, the two Soret bands at wavelengths 419 and 482 nm, along with well-pronounced Q-band at 700 nm, are present, which indicates that the TSPP compound forms *J*-aggregates in the porous silica film [41,42]. It was previously confirmed that nano-assembled thin films with TSPP in *J*-aggregate state are particularly sensitive to ammonia gas [42]. In this work, mesoporous SiO_2 NPs films infused with TSPP were to be used to detect the presence of ammonia in aqueous solutions. In order to study the stability of the PDDA/SiO_2 films infused with TSPP in aqueous solutions, they were immersed into water several times, with the resultant changes in absorption spectra shown in Figure 12. The second Soret-band (482 nm) and Q-band (at 700 nm) disappeared when the film was immersed into water, accompanied by a decrease in the absorbance of the first Soret-band (419 nm), which indicates the partial removal of the adsorbed TSPP molecules from the film [42]. We can speculate that this phenomenon is a result of the cleavage of the *J*-aggregates of TSPP formed in the space between the SiO_2 NPs.

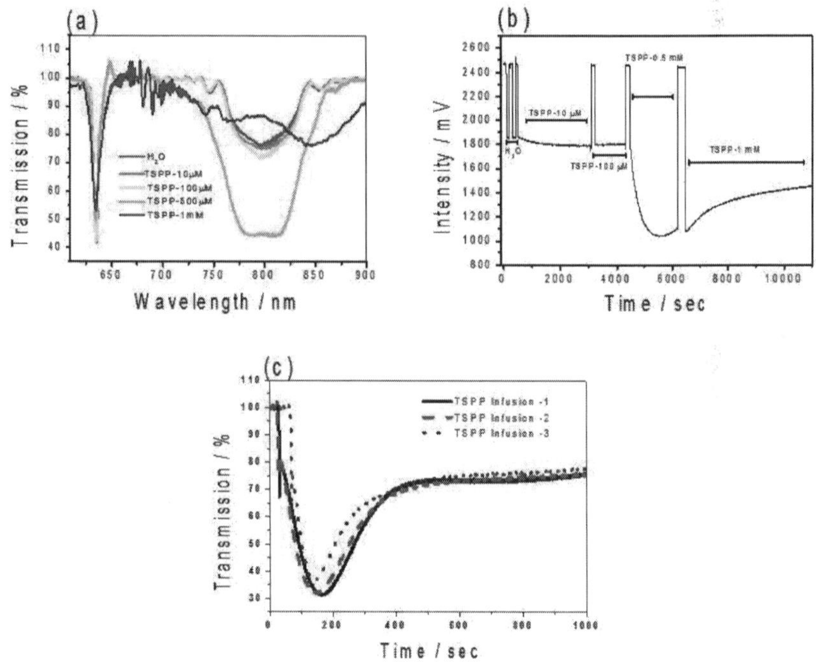

Figure 11. a) Transmission spectra of the $(PDDA/SiO_2)_{10}$ coated LPG; (b) transmission change recorded at 800 nm in response to different concentrations of TSPP (from 10 μM to 1 mM in water) and (c) dynamic response to the three infusions of TSPP into the PDDA/SiO_2 porous film recorded at 800 nm; the infusion was conducted after complete removal of TSPP from the PDDA/SiO_2 using an NH_3 solution of concentration 1000 ppm.

The observed absorbance decrease reached a steady state after several immersions, indicating that only strongly bound TSPP molecules remained in the PDDA/SiO_2 porous film. Subsequently, immersion of the PDDA/SiO_2 film into aqueous solutions containing ammonia led to the further removal of TSPP from the film, resulting in a decrease of the absorbance at 419 nm in proportion to the ammonia concentration, as shown in Figure 12. The absorbance at 419 nm almost disappeared, at an ammonia concentration of 170 ppm.

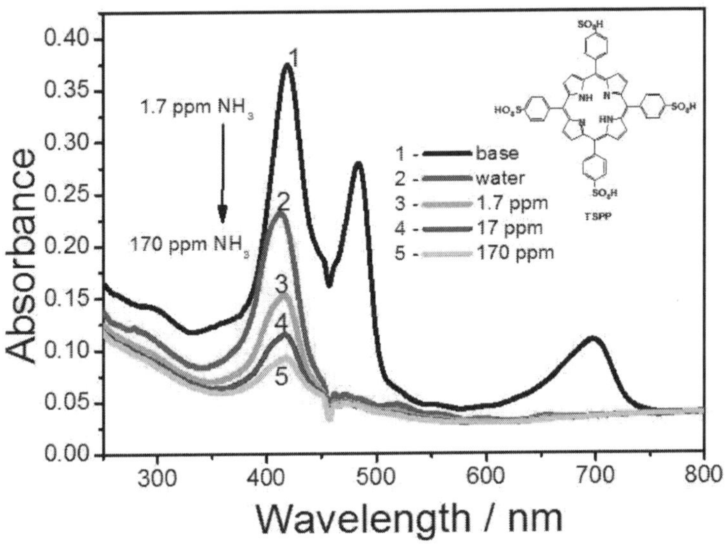

Figure 12. Absorption spectra of the $(PDDA/SiO_2)_{10}$ film infused with TSPP on a quartz plate: before immersion into H_2O (solid line) and after immersion into ammonia solutions of concentration from 0 ppm, 1.7 ppm, 17 ppm and 170 ppm (dashed lines); inset shows the structure of the TSPP.

The amount of TSPP infused inside the mesoporous film was estimated from QCM measurements to be 1135.6 ng (1.13 nmol). Consequently, considering the amount of adsorbed PDDA to be 250 ng (278 Hz for 10 cycles, 1.50 nmol as monomer unit of PDDA), the porous film contains sufficient binding sites for the infused TSPP molecules using the electrostatic interaction. When the QCM electrode that had been coated with a $(PDDA/SiO_2)_{10}$ film infused with TSPP was immersed into water and into ammonia, a similar trend to that measured using a UV-vis spectrophotometer was observed [36]. In particular, the frequency increased after each immersion into water (for 5 min) by up to a factor of 5, indicating the desorption of TSPP from the film; about 70% of the employed TSPP molecules were removed from the film [36]. Further immersion into water did not lead to a significant frequency change, suggesting that strongly bound TSPP (ca. 30% of the employed TSPP molecules) remained in the mesoporous film. When the film was exposed to ammonia solutions of different concentrations, further desorption of TSPP was observed and the total mass loss was 292 ng, indicating the almost complete removal of TSPP from the film [36]. Consequently, ca. 51 ng of TSPP remained in the film after ammonia treatment.

For ammonia detection, the sensing mechanism may be based upon changes in the RI of the coating resulting from chemically induced adsorption or desorption of the functional material to or from the mesoporous coating. Porphyrins have tetrapyrrole pigments [43] and their optical spectrum in the solid state is different to that in solution, due to the presence of strong π–π interactions [43]. Interactions with some chemical species can produce further spectral changes, thus creating the possibility that they can be used in the development of optical sensor systems. For instance, exposure of TSPP that has sulfonic functional groups to ammonia leads to the modification of the absorption spectrum [42].

5. AMMONIA SENSING

The sensitivity to ammonia in water of an LPG coated with a $(PDDA/SiO_2)_{10}$ film that was infused with TSPP was characterized by sequential immersion of the coated LPG into ammonia solutions with different concentrations (0.1, 1, 5 and 10 ppm). The lower ammonia concentrations were prepared by dilution of the stock solution of 28 wt%. In order to assess the stability of the base line, the coated LPG was immersed several times into 150 μL of pure water. The decrease of attenuation of the second resonance band, LP_{021}, at 800 nm, indicates the partial removal of the adsorbed TSPP molecules as discussed above. The equilibrium state was achieved after several exposures into water. For the ammonia detection, the LPG fibre was exposed into a 150 μL ammonia solution of 0.1 ppm, followed by drying and immersion into ammonia solutions of 1, 5 and 10 ppm.

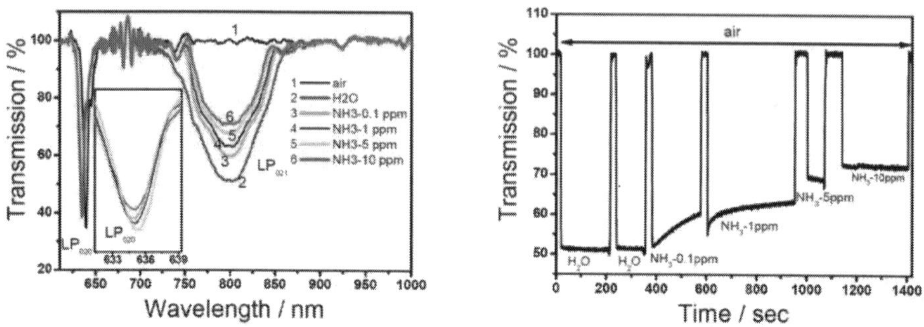

(a) (b)

Figure 13. a) Transmission spectra of the LPG coated with a TSPP infused $(PDDA/SiO_2)_{10}$ film due to immersion into water and into ammonia solutions of different concentrations: "H_2O", LPG exposed into water; "air", LPG in air after drying with N_2 gas; "NH_3 x ppm", LPG exposed into a x ppm ammonia solution, where x = 0.1, 1, 5 and 10. (b) Dynamic response to water and ammonia solutions (0.1, 1, 5 and 10 ppm) recorded at 800 nm; LP_{020} and LP_{021} are labelling the linear polarized 020 and 021 modes, respectively.

The response of the transmission spectrum to varying concentration of ammonia is shown in Figure 13a. The dynamic response of the sensor was assessed by monitoring the transmission at the centre of the LP_{021} resonance band at 800 nm. The response is shown in Figure 13b, where "air" region and "H_2O" and "NH_3" regions correspond to the transmission recorded at 800 nm after drying the LPG and exposing the devise into water and ammonium solutions, respectively. After repeating the process of immersion in water and drying 4 times, the recorded spectrum was stable, demonstrating the robustness and stability of the employed molecules in aqueous environments (H_2O regions indicated in Figure 13). On immersion in 1 ppm and 5 ppm ammonia solutions, the transmission measured at 800 nm increases. The transmission when the coated LPG was immersed in a 10 ppm ammonia solution exhibits a further increase, reaching a steady state within 100 s, as shown in Figure 13b. The resonance feature corresponding to coupling to the LP_{020} cladding mode exhibits additional small red shifts of 0.5 and 1.5 nm when subsequently immersed in solutions of 1 ppm and 10 ppm ammonia concentration, respectively, along with decreases in amplitude, as shown in Figure 13a. The limit of detection (LOD) for the 100 μm period LPG coated with a $(PDDA/SiO_2)_{10}$ film that was infused with TSPP was 0.14 ppm and 2.5 ppm when transmission and wavelength shift were measured respectively. The LOD was derived from the calibration curve and the following equation [36, 44].

$$\mathrm{LOD}=3\cdot\sigma/m \tag{2}$$

where σ is the standard uncertainty obtained as a symmetric rectangular probability [45]; *m* is the slope of the calibration curve. The sensing mechanism postulated is based upon the UV-vis spectrometer, QCM and LPG fibre measurements, and can be illustrated using Figure 14. As mentioned previously, while most of TSPP molecules form J-aggregates in the PDDA/SiO2 film, some of the molecules are present in monomeric forms, as shown in Figure 14a, which may be easily sulfonated into species with neutral ($TSPP^{4-}$) and protonated (H_2TSPP^{2-}) pyrrole rings in water. As can be concluded from UV-vis measurements, on immersion of the TSPP-infused $PDDA/SiO_2$ film into water, most of J-aggregated TSPP molecules are removed from the mesopores between SiO_2 NPs, Figure 12. This indicates that the intermolecular interaction between J-aggregates of TSPP can be easily broken in water. However, some strongly bond TSPP compounds remained in the porous film. Figure 14b shows the chemical reactions involved in the interaction of ammonia with the $TSPP^{4-}$ and H_2TSPP^{2-} monomers. The electrostatic interaction between TSPP and PDDA is disturbed by the formation of ammonium ions and this causes the further desorption of the TSPP compound from the $PDDA/SiO_2$ film and consequently a decrease in the refractive index of the film.

It is important to consider the influence of the pH of the ammonia solutions on the sensor response, as the electrostatic interaction between TSPP and PDDA can be disturbed by OH^- ions in solution. To check this, the pH of the ammonia solutions used in this work were measured using a compact pH meter (B-211, Horiba), showing 7.3, 7.3 and 7.6 for 0.1, 1, and 10 ppm solutions, respectively. Thus, the molar concentration of NH_3 and OH^- is estimated to be 0.58, 5.8 and 58 mM for NH_3 and 0.25, 0.25 and 50 nM for OH^-, respectively. These data reveal that the concentration of OH^- does not have a significant role and the sensing mechanism is mainly based on the basicity of ammonia, as shown in Figure 14b. The cross sensitivity of the LPG sensor was tested using ethanol and methanol aqueous solutions. There was no measurable response of the sensor at the concentration levels similar to those tested for ammonia (0.1, 1, 10 and 100 ppm) indicating high selectivity of the sensor device to ammonia over those analytes. At much higher concentrations (10,000 ppm), however, blue-shift of the LP_{020} band and decrease in transmission at 800 nm was registered, which can be ascribed to the change of the bulk RI of the solution [6].

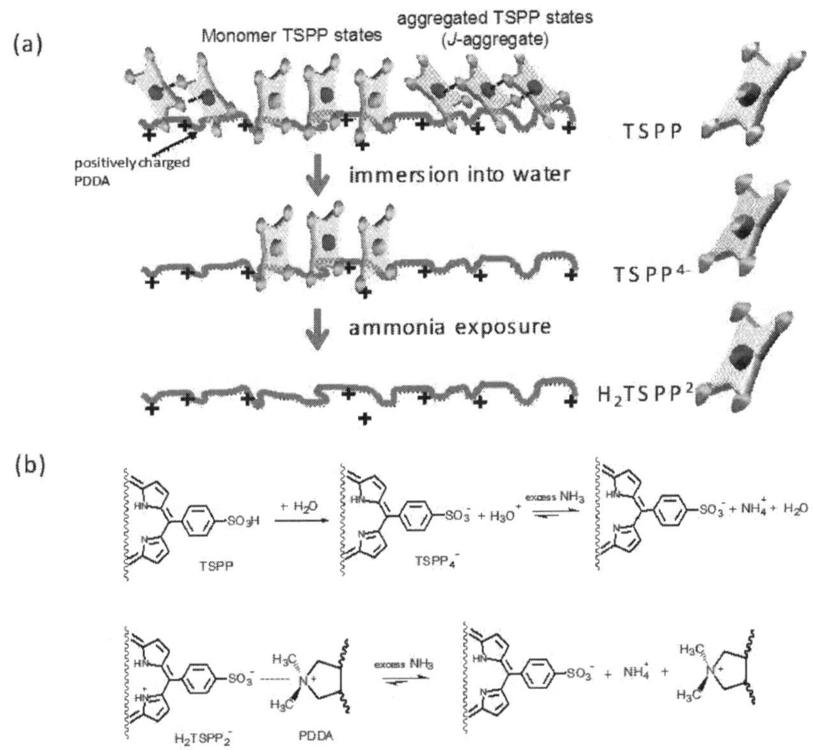

Figure 14. Schematic illustration of the sensing mechanism.

The assessment of the cross sensitivity to other amine containing compounds is in progress. When compared with other ammonia optical sensors [46] the developed LPG device shows similar detection levels to coulorometric [47] and absorption spectroscopy [48] devices. In particular, the current LPG optical fibre sensor modified with a mesoporous thin film offers unique advantages such as versatile chemical infusion of various chemicals into the mesopores, fast response time owing to the easy analyte penetration and robustness. In addition, due to unique properties of the optical fibre such as biocompatibility, multiplexing, small size, immunity to electromagnetic interference and possibility to work in harsh environment, a cost effective, portable sensor system can be produced. To further elucidate the sensing principle of the LPG sensor, the response of a $(PDDA/SiO_2)_{10}$ coated LPG and of a PAA-infused $(PDDA/SiO_2)_{10}$ coated LPG was further examined. There was no change when the $(PDDA/SiO_2)_{10}$ and PAA infused $(PDDA/SiO_2)_{10}$ coated LPGs were exposed to ammonia solutions which have similar values of RI, 1.3329, from 1 ppm to 100 ppm concentrations as measured using a portable refractometer (R-5000, Atago). Thus, no change in sensor signal was recorded in the $(PDDA/SiO_2)_{10}$ coated LPG. On the other hand, higher affinity to ammonia based on the acid-base interaction is expected in the PAA-infused $(PDDA/SiO_2)_{10}$ coated LPG [39]. However, no response to ammonia was observed. This indicates that the adsorption of ammonia in small molecular size results in a small RI change. Consequently, the selective desorption of the functional compound with a high RI from the mesoporous film results in a significant increase in the device sensitivity, as shown in Scheme 1. This measurement principle can be further explored for a different set of analyte-functional compound pair including biological analytes expanding the application range of the proposed LPG device.

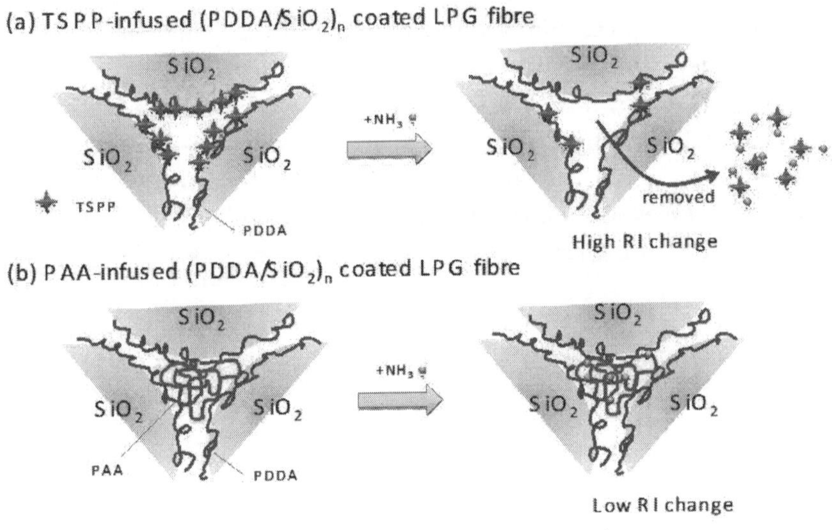

Scheme 1. Schematic illustration of the ammonia sensing mechanism for the LPG fibres modified with (a) TSPP-infused $PDDA/SiO_2$ and (b) PAA-infused $PDDA/SiO_2$ films.

6. DEVICE REPEATABILITY

It was observed that, if the sensor was repeatedly exposed to a certain concentration of ammonia following a washing and drying cycle, the response was not reproducible, in that on each subsequent exposure to ammonia the extinction of the band was further reduced, as shown in Figure 13. The extinction of the band still changes in time and the effect saturates with the increase of the ammonia concentration, in the way indicated in Figure 13 b. The linear dependence of the sensor response upon the ammonia concentration indicates that magnitude of the change on each exposure is the same; thus proposed device can be employed for measurements of the cumulative exposure to ammonia. However, after a number of repeated exposures, with the number being dependent on the concentration of the ammonia solution to which the device was exposed, the sensitivity was exhausted, and exposure to the ammonia solution would produce a spectrum equivalent to that obtained after deposition of the porous film, but before TSPP infusion [5]. It was found that the TSPP compound could be infused into the porous film again by the exposure of the LPG to a 1 mM solution of TSPP (step (vii) in Figure 5). The reproducibility of the device was tested by the exposure of the LPG coated with the TSPP infused $(PDDA/SiO_2)_{10}$ film to an ammonia solution of concentration >1000 ppm (concentration chosen to ensure that the TSPP was completely desorbed from the PDDA/SiO2 film), followed by washing with pure water and immersion in a 1 mM TSPP solution in order to regenerate the original film properties [5], Figure 11c. The procedure was repeated 5 times and the same behaviour and resulting LPG transmission spectrum were obtained after each reinfusion process. It should be noted that ammonia measurements were conducted 3 times at each concentration. These results indicate that highly reproducible measurements can be conducted by employing TSPP reinfusion step.

7. SUMMARY

The response of the transmission spectrum of an LPG of period 100 μm to the deposition of a multilayer film of SiO_2 NPs and the subsequent infusion of a porphyrin into the porous coating has been characterized. The infusion of the functional materials, chosen to be sensitive to the analyte of interest, into the base mesoporous coating was reported. Two possible sensing mechanisms have been exploited, based upon changes in the refractive index of the coating resulting from (1) chemically induced RI changes of the mesoporous coating at the adsorption of the analyte to the functional material, namely PAA and (2) chemically induced desorption of the functional material, namely TSPP, from the mesoporous coating. The operation of the device as a re-useable ammonia sensor with a minimum detection level of 0.14 ppm and a response time of approximately 100 s exploiting desorption sensing mechanism has been reported. On the other hand, the ammonia adsorption to the carboxylic functional groups of the PAA resulted in a small RI change and low sensitivity to analyte. The film thicknesses and functional material infusion time employed in this work have been determined empirically, as the calculation of the RI of the porous coating infused with TSPP and immersed in water would be highly complex. Operation of the system at the point of coincidence of the mode transition region and phase matching turning point for films of larger/smaller thickness could be achieved by decreasing/increasing the quantity of the function dye infused into the film, and this may influence the minimum detectable concentration and sensitivity. Such issues are currently under investigation. Advantages of the proposed method lie in the ability to control functionality of the coating by, for instance, choosing different matrix polymers that extend the class of the detectable analytes. Additionally, infusion of different types of functional compounds would allow the detection of different chemicals using a similar principle of operation. Future work is planned to demonstrate the current system for sensing a variety of chemical and biological compounds and for gas sensing.

ACKNOWLEDGEMENT

This work was supported by the Regional Innovation Cluster Program of the Ministry of Education, Culture, Sports, Science and Technology (MEXT), Japan and partly by the Ministry of Knowledge Economy (MKE, Republic of Korea) via the Fundamental R&D Program for Core Technology of Materials. The authors from Cranfield are grateful to the Engineering and Physical Sciences Research Council, EPSRC, UK for funding under grants EP/D506654/1 and GR/T09149/01.

REFERENCES

1. H. Tsuda, K. Urabe, Optical Fibre Long-Period Gratings Functionalised with Nano-Assembled Thin Films: Approaches to Chemical SensingSensors 200994559

2. T. M. Libish, J. Linesh, P. Biswas, S. Bandyopadhyay, K. Dasgupta, P. Radhakrishnan, Optic. Fiber, Period. Long, Based. Grating, for. Sensor, Oil. Coconut, Detection. Adulteration, Sensors & Transducers Journal 2010114102

3. James S W, Tatam R P.Optical Fibre Long-Period Gratings Functionalised with Nano-Assembled Thin Films: Approaches to Chemical SensingMeasurement Science and Technology 2003R49R61.

4. A. Martinez-Rios, D. Monzon-Hernandez, I. Torres-Gomez, Highly sensitive cladding-etched arc-induced long-period fiber gratings for refractive index sensing. Optics Communications 2010283958

5. S. Korposh, S. W. James, S. Lee-W, S. M. Topliss, S. C. Cheung, W. J. Batty, R. P. Tatam, Optical Fibre Long-Period Gratings Functionalised with Nano-Assembled Thin Films: Approaches to Chemical SensingOptics Express 20101813227

6. S. Korposh, S. W. James, R. P. S. Tatam-W, Lee, Optical Fibre Long-Period Gratings Functionalised with Nano-Assembled Thin Films: Approaches to Chemical SensingTech. 2011s.

7. K. Schroeder, W. Ecke, R. Mueller, R. Willsch, A. A. A. Andreev, Bragg. fibre, refractometer. grating, Measurement Science and Technology 200112757

8. A. Asseh, S. Sandgren, H. Åhlfeldt, B. Sahlgren, R. Stubbe, G. Edwall, Fiber optical bragg grating refractometer. Fiber and Integrated Optics 19985162

9. A. Iadicicco, S. Campopiano, A. Cutolo, M. Giordano, A. Cusano, Optical Fibre Long-Period Gratings Functionalised with Nano-Assembled Thin Films: Approaches to Chemical SensingSensors and Actuators, B: Chemical 2006120231

10. G. Laffon, P. Ferdinand, Tilted short-period fibre-Bragg-grating-induced coupling to cladding modes for accurate refractometry. Measurement Science and Technology 200112765

11. S. J. Buggy, E. Chehura, S. W. James, R. P. Tatam, Optical Fibre Long-Period Gratings Functionalised with Nano-Assembled Thin Films: Approaches to Chemical SensingJournal of Optics A: Pure and Applied Optics 2007S60S65.

12. D. Paladino, A. Cusano, P. Pilla, S. Campopiano, C. Caucheteur, P. Mégret, Spectral behavior in nano-coated tilted fiber Bragg gratings: Effect of thickness and external refractive index. IEEE Photonics Technology Letters 2007192051

13. C. Caucheteur, D. Paladino, P. Pilla, A. Cutolo, S. Campopiano, M. Giordano, A. Cusano, P. Mégret, External refractive index sensitivity of weakly tilted fiber Bragg gratings with different coating thicknesses. IEEE Sensors Journal 200881330

14. V. Bhatia V, Vengsarkar A M.Optical fibre long-period grating sensors. Optics Letters 199621692

15. Q. Shi, B. T. Kuhlmey, Optical Fibre Long-Period Gratings Functionalised with Nano-Assembled Thin Films: Approaches to Chemical SensingOptics Communications20092824723

16. E. Davies, R. Viitala, M. Salomäki, S. Areva, L. Zhang, I. Bennion, Optical Fibre Long-Period Gratings Functionalised with Nano-Assembled Thin Films: Approaches to Chemical SensingJournal of OpticsA: Pure and Applied Optics 2009art. 015501

17. T. Allsop, D. J. Webb, I. A. Bennion, of. comparison, sensing. the, of. characteristics, period. long, written. gratings, three. in, types. different, fiber. of, Optical Fiber Technology 20039210

18. X. Yu, P. Shum, G. B. Ren, N. Q. Ngo, Optical Fibre Long-Period Gratings Functionalised with Nano-Assembled Thin Films: Approaches to Chemical SensingOptics Communications20082814555

19. Y. Zhu, Z. He, J. Kaňka, H. Du, Numerical analysis of refractive index sensitivity of long-period gratings in photonic crystal fiber. Sensors and Actuators, B: Chemical 200812999

20. Q. Shi, B. T. Kuhlmey, Optical Fibre Long-Period Gratings Functionalised with Nano-Assembled Thin Films: Approaches to Chemical SensingOptics Communications20092824723

21. Rees N D, James S W, Tatam R P, Ashwell G J.Optical fiber long-period gratings with Langmuir- Blodgett thin-film overlays. Optics Letters 200227686

22. T. Allsop, F. Floreani, K. P. Jedrzejewski, P. V. S. Marques, R. Romero, D. J. Webb, I. Bennion, Spectral characteristics of tapered LPG device as a sensing element for refractive index and temperature. Journal of Lightwave Technology 200624870

23. A. Iadicicco, S. Campopiano, M. Giordano, A. Cusano, Optical Fibre Long-Period Gratings Functionalised with Nano-Assembled Thin Films: Approaches to Chemical SensingApplied Optics2007466945

24. A. Martinez-Rios, D. Monzon-Hernandez, I. Torres-Gomez, Optical Fibre Long-Period Gratings Functionalised with Nano-Assembled Thin Films: Approaches to Chemical SensingOptics Communications 2010283958

25. I. M. Ishaq, A. Quintela, S. W. James, G. J. Ashwell, J. M. Lopez-Higuera, R. P. Tatam, Modification of the refractive index response of long period gratings using thin film overlays. Sensors and Actuators, B: Chemical 2005107738

26. M. Smietana, M. L. Korwin-Pawlowski, W. J. Bock, G. R. Pickrell, J. Szmidt, Optical Fibre Long-Period Gratings Functionalised with Nano-Assembled Thin Films: Approaches to Chemical SensingMeasurement Science and Technology 2008art. 085301

27. Rees N D, James S W, Tatam R P, Ashwell G J.Optical fiber long-period gratings with Langmuir- Blodgett thin-film overlays. Optics Letters 200227686

28. A. Cusano, A. Iadicicco, P. Pilla, L. Contessa, S. Campopiano, A. Cutolo, M. Giordano, Optical Fibre Long-Period Gratings Functionalised with Nano-Assembled Thin Films: Approaches to Chemical SensingOpt. Lett. 2005302536

29. J. Yang, L. Yang, C. Xu-Q, Y. Li, Optical Fibre Long-Period Gratings Functionalised with Nano-Assembled Thin Films: Approaches to Chemical SensingJournal of Lightwave Technology200725372

30. I. Del Villar, M. Achaerandio, I. R. Matias, F. J. Arregui, Optical Fibre Long-Period Gratings Functionalised with Nano-Assembled Thin Films: Approaches to Chemical SensingOpt. Lett. 200530720

31. I. Del Villar, I. R. Matias, F. J. Arregui, Optical Fibre Long-Period Gratings Functionalised with Nano-Assembled Thin Films: Approaches to Chemical SensingJ. Opt. Soc. Am. A 200623651

32. Cheung S C, Topliss S M, James S W, Tatam R P.Optical Fibre Long-Period Gratings Functionalised with Nano-Assembled Thin Films: Approaches to Chemical SensingJ. Opt. Soc. Am. B 200825897

33. J. M. Corres, I. R. Matias, I. del Villar, F. J. Arregui, of. p. H. Design, in. sensors, fiber. long-period, using. gratings, nanocoatings. I. E. E. polymeric, IEEE Sens. J. 20077455

34. J. Keith, L. C. Hess, W. U. Spendel, J. A. Cox, G. E. Pacey, Optical Fibre Long-Period Gratings Functionalised with Nano-Assembled Thin Films: Approaches to Chemical SensingTatanta 200670818

35. Schick G A, Schreiman I C, Wagner R W, Lindsey J S, Bocian D F.Optical Fibre Long-Period Gratings Functionalised with Nano-Assembled Thin Films: Approaches to Chemical SensingJ. Am. Chem. Soc. 19891111344

36. S. Korposh, R. Selyanchyn, W. Yasukochi, S. Lee-W, S. James, R. Tatam, Optical Fibre Long-Period Gratings Functionalised with Nano-Assembled Thin Films: Approaches to Chemical SensingMaterials Chemistry and Physics2012133784

37. S. Lee-W, N. Takahara, S. Korposh, D. Yang-H, K. Toko, T. Kunitake, Optical Fibre Long-Period Gratings Functionalised with Nano-Assembled Thin Films: Approaches to Chemical Sensing2010822228

38. Ye C C, James S W, Tatam R P.Simultaneous temperature and bend sensing with long-period fiber gratings. Optics Letters 2000251007

39. J. Bravo, L. Zhai, Z. Wu, R. E. Cohen, M. F. Rubner, Transparent superhydrophobic films based on silica nanoparticles. Langmuir 2007237293

40. E. P. K. Currie, A. B. Sieval, G. J. Fleer, Stuart. M. A. Cohen, Polyacrylic acid brushes: Surface pressure and salt-induced swelling. Langmuir 2000168324

41. S. Korposh, S. Kodaira, W. J. Batty, S. W. James, S. Lee-W, Nano-assembled thin film gas sensor. II. An intrinsic high sensitive fibre optic sensor for ammonia detection. Sensor and Materials. 20092141790189

42. S. O. Korposh, N. Takahara, J. J. Ramsden, S. Lee-W, T. Kunitake, Nano-assembled thin film gas sensors. I. Ammonia detection by a porphyrin-based multilayer film. J. Biol. Phys. Chem. 20066125

43. K. M. Kadis, K. M. Smith, R. Guilard, Porphyrin. The, vol. Handbook, 112 , , Academic, Press, 2003

44. Swartz ME, Krull IS.Optical Fibre Long-Period Gratings Functionalised with Nano-Assembled Thin Films: Approaches to Chemical SensingMarcel Dekker, Inc.: NY USA; 1997

45. G. Rafael, C. Possetti, R. C. Kamikawachi, M. Muller, J. L. Fabris, Optical Fibre Long-Period Gratings Functionalised with Nano-Assembled Thin Films: Approaches to Chemical SensingJ. Lightwave Technology (OFS-21). (2011DOIJLT.2011.2167500.

46. B. Timmer, W. Olthuis, A. Berg, Optical Fibre Long-Period Gratings Functionalised with Nano-Assembled Thin Films: Approaches to Chemical SensingSens. Act. B. 2005107666

47. A. Yimit, K. Itoh, M. Murabayashi, Optical Fibre Long-Period Gratings Functionalised with Nano-Assembled Thin Films: Approaches to Chemical SensingSens. Act. B. 200388239

48. H. Huszár, A. Pogány, Z. Bozóki, Á. Mohácsi, L. Horváth, G. Szabó, Optical Fibre Long-Period Gratings Functionalised with Nano-Assembled Thin Films: Approaches to Chemical SensingSens. Act. B. 20081341027

CHAPTER 8

Fiber Bragg Grating Technology for New Generation Optical Access Systems

Oskars Ozolins, Vjaceslavs Bobrovs, Jurgis Porins and Girts Ivanovs

[1] Riga Technical University, Telecommunications Institute, Latvia

1. INTRODUCTION

Thanks to achievements in photonics technologies optical networks have gained a notably increase in capacity per single fiber during the last decade [1]. Demand for greater transmission speed has been increasing exponentially because of the impulsive spread of Internet services (see Fig. 1) [2]. At the same time, the radical improvement of the capability of digital technologies has made feasible expanding multimedia services [1]. Therefore bandwidth intensive applications and exponential Internet traffic growth are continuing to drive the further penetration of optical fiber into the optical access systems [2].

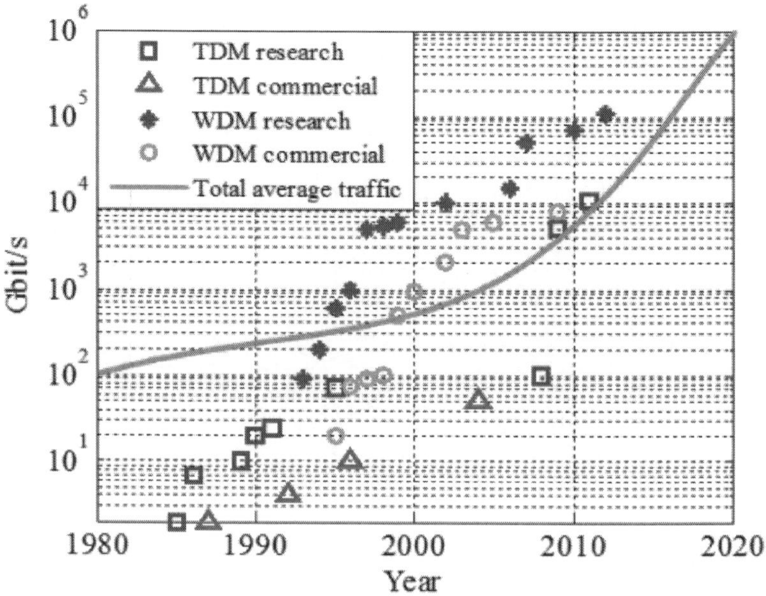

Figure 1. System capacity (per fiber) and network traffic [3]

A new generation access network becomes indispensable to upgrade the systems [4, 5]. The passive optical network (PON) has the feasibility to lead the deployment of new high-capacity and future-proof broadband networks [6]. The use of wavelength division multiplexing (WDM) in the access networks is further defensible in terms of video services support. WDM solutions are forging ahead towards higher data transmission rate and lower channel spacing to utilize available bandwidth more effectively [7].

The main reasons behind the implementation of new generation systems are to meet demand of capacity, user density requirements and scalability, while ensuring that the cost per unit bandwidth is lowest possible [3, 8]. The novel concept is WDM-direct in which multiple wavelengths are directly connected to each optical network unit (ONU) [9, 10]. Increasing spectral efficiency is important for building efficient WDM-direct systems, since this allows the optical infrastructure to be shared among many channels, and thus reduces the cost per transmitted information bit in a fully loaded system [11]. High performance optical filters are groundwork for realization of high speed dense WDM (DWDM)-direct systems where coherent and incoherent crosstalk between adjacent channels becomes a main source of degradation:

adjacent channels interfere with each other upon detection, and the resulting beating gives rise to signal distortions, provided that the beat frequencies lie within the bandwidth of the detection electronics [12, 11].

Proposed approach for increasing the transmission capacity is to reduce the channel spacing of a DWDM-direct system to the minimum while keeping the mature and well developed optical filter technologies like fiber Bragg grating (FBG). To realize proposed approach limitating factors must be taken into account. One part of these factors is related to efficient bandwidth of FBG which in proposed approach is determined employing optical signals (transmission speed 2.5 Gbit/s and 10.3125 Gbit/s which conform 2 Gigabit Ethernet (GE) and 10 GE of Ethernet hierarchy) with different wavelength offset value within filter pass-band. Other part of factors is related to evaluation of the minimal channel spacing for concrete FBG in DWDM-direct system.

2. OPTIMAL COMPLEX TRANFER FUNCTION FOR ACCESS SYSTEMS

A FBG is periodic variation of the refractive index along the propagation direction in the core of optical fiber that reflects particular wavelengths of light and transmits all others. Low channel spacing and high data transmission rate sets strict requirements for DWDM filter characteristics and any imperfections in their parameters, such as amplitude and phase responses, becomes critical. Understanding and distracting of those optical filter imperfections to high speed DWDM-direct systems are of great importance [7].

Low channel isolation from adjacent channels is one of these imperfections in optical filter parameters. To ensure high channel isolation we need to inscribe FBG filters with complex apodization profiles. Changes in apodization profiles emerge in different filter bandwidth at -3 dB and -20 dB level and suppression of undesirable side lobes in optical filter amplitude response. Performance of three different apodization profiles and their influence on DWDM-direct systems main parameters: channel spacing and data transmission rate have been evaluated numerically.

2.1. Simulation method

To numerically evaluate impact of different FBG apodization profiles on high speed DWDM-direct system combination of two different simulation programs was used: Bragg Grating Filters Synthesis 2.6 (BGFS 2.6) simulation program for mathematical description of FBG optical filter and OptSim 5.2 simulation program to simulate high speed DWDM-direct systems. In the BGFS 2.6 simulation program different FBG optical filters with defined apodization profiles were realized. This simulation program is based on Transfer Matrix Method (TMM). TMM is used to create a numerical periodic non-uniform FBG filters. It is applied to solve the coupled mode equations and to obtain the spectral response of the fiber Bragg grating. In this approach, the grating is divided into uniform sections. Each section is represented by a 2x2 matrix. By multiplying these matrices, a global matrix that describes the whole grating is obtained (see Fig. 2. and equation 1):

$$\begin{cases} T_M = T_1 \cdot T_2 \cdot T_3 \cdot \ldots \cdot T_i, \\ \begin{bmatrix} R_M \\ S_M \end{bmatrix} = T_M \cdot \begin{bmatrix} R_0 \\ S_0 \end{bmatrix}. \end{cases} \tag{1}$$

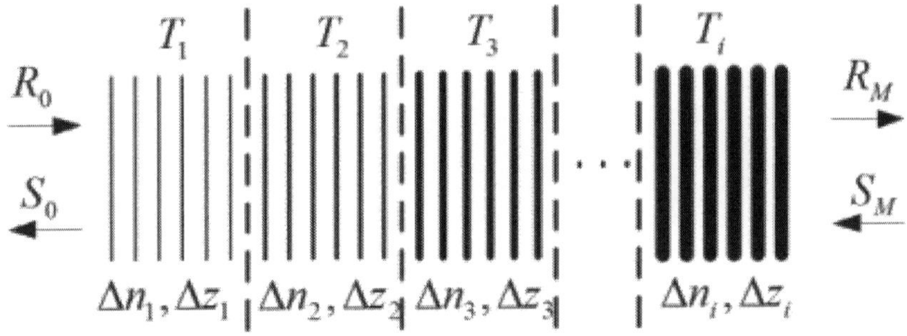

Figure 2. Transfer matrix method used to obtain the spectral characteristics of a fiber Bragg grating (E(r,ϕ,z)=r^Er(r,ϕ,z)+ϕ^Eϕ(r,ϕ,z)+z^Ez(r,ϕ,z). - average index of refraction, (∇2+*k*2){*EzHz*}=0,- section length, ∇2=∂2∂*r*2+1*r*∂∂*r*+1*r*2∂2∂*ϕ*2+∂2∂*z*2,- electromagnetic waves amplitudes)

The reflection coefficient of the entire grating is defined as:

$$\rho = \frac{S_0}{R_0}. \tag{2}$$

The main drawback of this method is that *M* may not be made arbitrarily large, since the coupled-mode theory approximations are not valid when uniform grating section is only a few grating periods long. Thus, it requires $\nabla \times E = -j\omega\mu H, \nabla \times H = j\omega\varepsilon E$, >> T [13, 14].

OptSim 5.2 simulation program uses method of calculation that is based on solving a complex set of differential equations, taking into account optical and electrical noise, linear and nonlinear effects. Two ways of calculation are possible: Frequency Domain Split Step (FDSS) and Time Domain Split Step (TDSS) methods. These methods differ in linear operator L calculations: FDSS does it in frequency domain, but TDSS calculates linear operator in the time domain by calculating the convolution product in sampled time. The first method is easy to realize, but it may cause severe errors during simulation. In our simulation we used the second method, TDSS, which despite its complexity grants a precise result. The Split Step method is used in all commercial simulation tools to perform the integration of the fiber propagation equation:

$$\frac{\partial A(t,z)}{\partial z} = \{L + N\} A(t,z), \tag{3}$$

where $A(t,z)$ -the optical field; *L*-linear operator that stands for dispersion and other linear effects; *N* – operator that is responsible for all nonlinear effects. The idea is to calculate the equation over small spans of fiber Δz by including either linear or nonlinear operator. For instance, on the first span Δz only linear effects are considered, on the second – only nonlinear, on the third – again only linear [15]. Us it is noticed before, in numerical investigation are used two simulation programs: BGFS 2.6 – to realize FBG filters amplitude and phase responses and OptSim 5.2 to numerically evaluate high speed DWDM-direct systems. Realized FBG filter parameters were recorded in data file, which after simple mathematical calculations were used in OptSim 5.2 simulation program to build user defined optical filters.

1.2. Simulation scheme and results

Simulation scheme (see Fig.3.) consists of transmitter, transmission line and receiver. Number of channels is chosen to evaluate influence of nonlinear optical effects (NOE): self – phase modulation (SPM), cross – phase modulation (XPM), four – wave – mixing (FWM) to used optical filters performance.

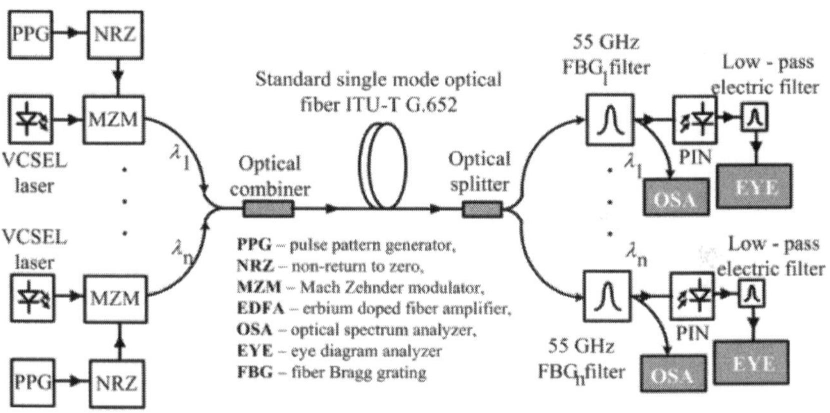

Figure 3. Simulation scheme for DWDM-direct transmission realization with FBG optical filters

The transmitter consists of pseudo-random data source with 2^{31}-1 bit sequence, non return to zero (NRZ) code former, continuous wavelength (CW) laser source and $LiNbO_3$-based external Mach Zehnder modulator (MZM). The data source produces a pseudo-random electrical signal, which represents the information we want to transmit via optical fibre. Then we use a code former to form NRZ code from incoming pseudo-random bit sequence. The NRZ has long been the dominant code format in fibre optical transmission systems, because of a relatively low electrical bandwidth for the transmitters and receivers and its insensitivity to the laser phase noise [16]. The optical pulses are obtained by modulating CW laser irradiation in MZM with previously mentioned bit sequence. Then formed optical pulses are sent directly to a different length standard single mode fibre (SSMF). The utilized fibre has a large core effective area 80 μm^2, attenuation $\alpha = 0.2$ dB/km, nonlinear refractive coefficient $n_k = 2.5\ 10^{-20}$ cm/W and dispersion 16 ps/nm/km at the reference wavelength $\lambda = 1550$ nm. Receiver block consists of optical filter, PIN photodiode (typical sensitivity -17 dBm) and Bessel – Thomson electrical filter (4 poles, 7.5 GHz -3dB bandwidth). To simulate insertion loss (polarization dependent loss: 0.1 dB, ripple insertion loss: 0.2 dB, splice and connector loss: 0.1dB) of optical filter we used optical attenuator.

The main idea of our simulations is to demonstrate FBG filters with different apodization profiles (see Fig. 4.) influence on high speed dense WDM communication systems. Investigation of high performance optical band-pass filters are groundwork for realization of high speed dense WDM communication systems.

The main problem is to ensure high channel isolation between adjacent channels. To realize channel isolation performance evaluation of FBG optical filter we used eye diagram, bit error rate (BER) and optical signal spectrum in different system configurations (different channel spacing and data transmission speed). We have chosen three different apodization profiles (see Fig.4.): rectangular, cosinusoidal and Gaussian, four channel spacing values: 200 GHz, 100 GHz, 50 GHz and 25 GHz and two data transmission speeds: 2.5 Gbit/s and 10. 3125 Gbit/s.

The results of BER dependence on channel spacing using FBG with rectangular, cosinusoidal and Gaussian apodization profiles are presented in Fig. 5. As we can see systems with 2.5 Gbit/s data transmission speed performance is better (BER values are lower) than systems with 10. 3125 Gbit/s data transmission speed. This can be explained by greater influence of chromatic dispersion on higher data transmission speed optical pulses. In addition, from results we can see that BER values are higher at 25 GHz channel spacing due to greater NOE influence and crosstalk. At both data transmission speeds and all channel spacing values, the worst performance showed the FBG optical filter with rectangular apodization profile. This is mainly because of great undesirable side lobes in optical filter amplitude response. These imperfections in filter amplitude response reduced channel isolation.

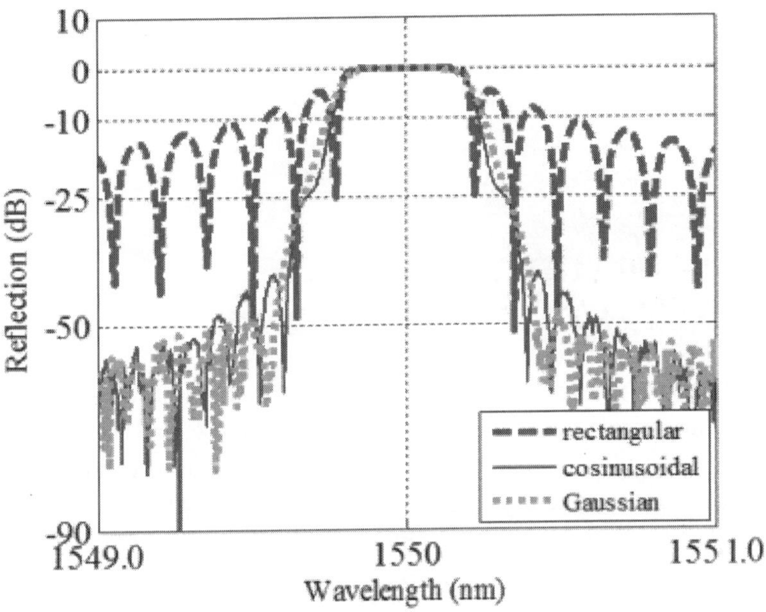

Figure 4. Amplitude response of 55 GHz FBG optical filters with different apodization profiles shown in inset

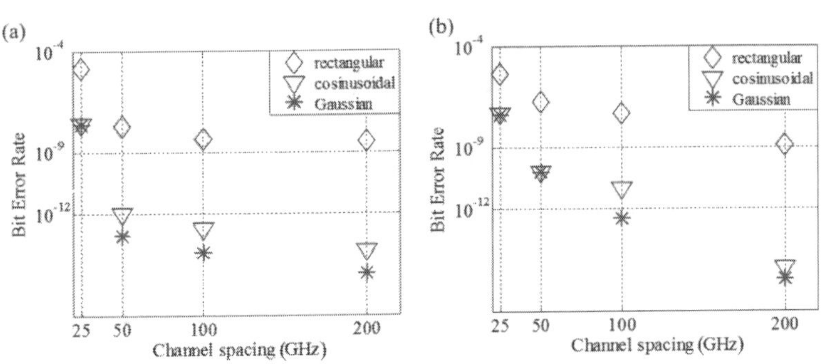

Figure 5. BER dependence on channel spacing using FBG with rectangular, cosinusoidal and Gaussian apodization profiles: a – 2.5 Gbit/s; b – 10.3125 Gbit/s data transmission speed. BER values measured at the worst channel.

As one could see from simulation results the influence of FBG filters with different apodization profiles on high speed DWDM-direct systems is enormous. To ensure high channel isolation and thus realize systems high performance FBG filter must be used with cosinusoidal or Gaussian apodization function because of narrow bandwidth at -20 dB level. FBG with rectangular apodization profile showed the worst performance which resulted in whole system degradation because of imperfections (great side lobes) in amplitude response.

2. FIBER BRAGG GRATING CHARACTERIZATION METHODS

Optical filters in optical transmission systems are a special subgroup of physical components defined in such a way that they select or modify parts of the spectrum of the signal [17, 18]. Signals and physical components can be expressed mathematically by complex functions describing amplitude and phase [17]. Amplitude transfer function describes loss dependency as function of wavelength for optical filter, but phase transfer function is responsible for introduced wavelength dependent amount of delay. There are different methods for transfer funtion characterization. Focus is related to the techniques which are for evaluation of phase transfer funtion and its related parematers.

2.1. Jones matrix eigenanalysis

Polarization mode dispersion (PMD) decreases transmission systems bandwidth and is a fundamental parameter of both: optical fiber and passive optical components. The difference between group delays for two principle states of polarization (PSP) is the differential group delay (DGD) [19]. The PMD value is the average of DGD values. DGD varies randomly with wavelength and time which stands for dispersive effects. Moreover a second-order effect of PMD can lead to optical pulses length changes [20]. As a consequence an optical component with birefringence devoid of chromatic dispersion (CD), can exhibit optical pulses length changes owing to the second-order effects of PMD [21]. PMD is characterized by a Jones Matrix as a function of wavelength [22]. Jones Matrix is common to represent the polarization state of an optical signal or the transfer matrix of a passive optical device. The transfer matrix of an optical device specifies the relationship between the input and output Jones vectors of the optical signal [20]. This transfer matrix can be characterized by measuring three output Jones vectors in response to three known input Jones vectors. It is worthwhile to note that although it is measured with a few specific polarization states of the optical signal, a Jones Matrix describes a passive optical device such as an optical filter and is independent of the input launching condition of the optical signal [22]. Two basic equations for this DGD estimation are given below. Thus, the two Eigen-values can be calculated from the products of Jones Matrix employing (4).

$$\rho_{1,2} = \frac{m_{11} + m_{22}}{2} \pm \sqrt{\left(\frac{m_{11} + m_{22}}{2}\right)^2 - m_{11} \cdot m_{22} + m_{12} \cdot m_{21}} \qquad (4)$$

Where $\rho_{1,2}$ is Eigen-values and $m_{11}, m_{12}, m_{21}, m_{22}$ - products of Jones Matrix [22]. Therefore DGD can be expressed as the group delay difference as the function of the optical frequency:

$$\Delta\tau(\omega) = \left|\tau_{g1} - \tau_{g2}\right| = \left|\frac{Arg(\rho_1 / \rho_2)}{\Delta\omega}\right| \qquad (5)$$

Where, τ_{g1} and τ_{g2} is group delays for two principle states of polarization,

$Arg(\rho_1 / \rho_2)$ denotes the phase angle of ρ_1 / ρ_2 [22]. Equation (5) shows the principle of Jones Matrix technique for DGD measurement. An important practical issue is to choose the frequency step size $\Delta\omega$ for the measurement. For a small step size the measurement would take a long time and instability of devices would have strong impact on the results. For a big step size the output optical signal state of polarization would rotate for more than 45° over each frequency step which leads to inaccurate results [19].

The Jones Matrix technique has a number of advantages compared to other DGD measurement methods [21]. First, it needs only a small wavelength window to perform a measurement. From this point of view method is more suitable for evaluation of detailed wavelength dependency of DGD and PMD. Second, Jones Matrix measurement can be made fast using automated procedures of polarization controller and a polarimeter. Third, the accuracy of the Jones Matrix technique is considered the best compared to other techniques [22]. Therefore Jones matrix Eigen-analysis (JME) is a measurement technique for accurately measure the DGD and PMD of any passive optical component [19].

In this research N7788B component analyser of Agilent Technologies was utilized to perform measurements of FBG with 55 GHz full width half maximum (FWHM) bandwidth. This technology is based on the JME which is the standard method for measuring DGD and PMD of passive optical devices [23].

Measurement scheme (see Fig.6) consists of $LiNbO_3$ polarization controller, polarimeter, tunable laser source and device under test (DUT): FBG with 55 GHz FWHM bandwidth.

Fig.7 shows measured amplitude transfer function and DGD as function of wavelength for FBG with 55 GHz FWHM bandwidth. Results show that DGD value within filters pass-band does not exceed 70 picoseconds.

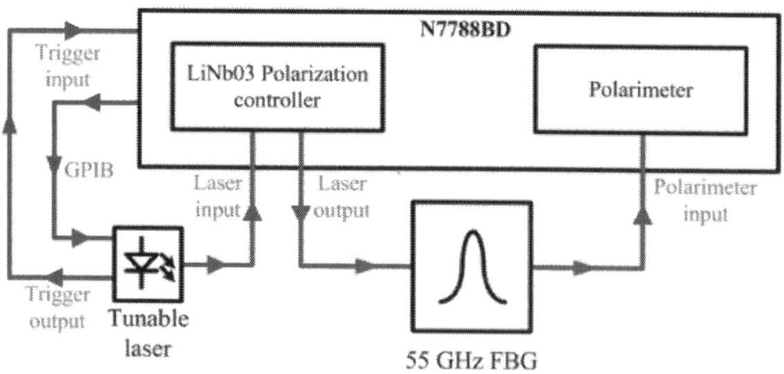

Figure 6. DGD measurement scheme [24]

DGD and PMD value increases at the edges of the optical filter amplitude transfer function which could be a degradation factor for optical signal transmitted through device [25]. Therefore the relative bandwidth available to each broadband access systems channel is reduced, meaning that the channel experiences the effect of the edge of the pass-band of the filter transfer function, where the dispersion effects is expected to be most significant [17].

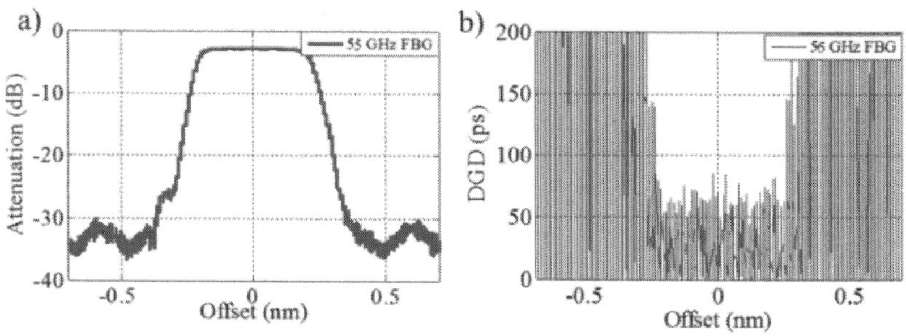

Figure 7. Measured amplitude transfer (a) function and differential group delay (b) of 55 GHz FBG

2.2. Modulation phase shift method

Agilent Technologies 86038B photonic dispersion and loss analyzer was employed for 55 GHz FBG filter parameters evaluation and numerous parameters were obtained: attenuation, group delay (GD) and chromatic dispersion as functions of wavelength. Test equipment is based on the modulation phase shift (MPS) method. In the conventional MPS method, light from a sinusoidal source is intensity modulated before being launched into the device under test [24]. MPS method obtains the group delay response of a device under test by measuring the change in phase of a sinusoidal radio frequency (RF) modulation envelope as the wavelength is changed [23].

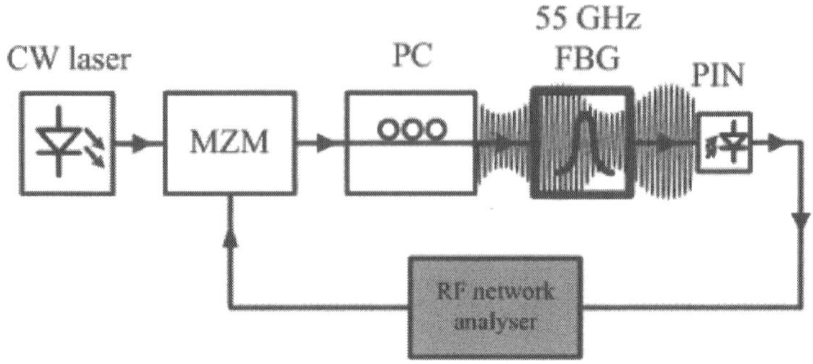

Figure 8. MPS method scheme for 55 GHz FBG measurements

See MPS method realization in Fig.8. The optical light source is a tunable distributed feedback laser. Light from a laser is sent to external MZM and is amplitude modulated (typically in the 100 MHz to 1.25 GHz range). After propagating through the DUT, the transmitted signal is detected by a PIN photodiode. A RF network analyzer is employed to provide a modulating signal of frequency f_m and to measure electrical phase difference between input and output signals [26, 27]. In practice, the wavelength is swept and the change in the group delay $\Delta\tau$ for each wavelength increment is calculated from the measured change in phase according to (6):

$$\Delta\tau_{(\Delta\lambda)} = -\frac{\Delta\varphi}{360^o}\cdot\frac{1}{f_m}. \tag{6}$$

Where the first factor is defined as the fractional cycle of RF phase shift and the second factor represents the period of the RF signal. The subscript *σ(z)=ωnco2δn¯¯¯¯¯¯cof∞−∞∫∞−∞et(x,y)·e∗t(x,y)dxdyκ(z)=ωncov4δn¯¯¯¯¯¯co∫∞−∞∫∞−∞Δε(x,y,z)et(x,y)·e∗t(x,y)dxdy,*

indicates that the change in group delay being measured was produced in response to an incremental change in wavelength. In (6) we can notice how the group delay and the measured electrical phase present opposite slopes [20, 27].

The attribute called dispersion is defined by:

$$D = \frac{\Delta\tau}{\Delta\lambda} \tag{7}$$

Where $\Delta\tau$ is the change in group delay in seconds corresponding to a change in wavelength $\Delta\lambda$ in meters. In real world applications, the dispersion parameter is given in units of picoseconds per nanometer. Combining (6) and (7), we obtain:

$$\Delta\varphi = -360^{o} \cdot D \cdot \Delta\lambda \cdot f_m \tag{8}$$

Equation (8) specifies that the amount of phase change obtained in response to a wavelength step is the product of total device dispersion, wavelength step and modulation frequency. This equation provides several key insights into the capabilities of the MPS measurement method. In order to achieve accurate measures it is important to have a stable wavelength step size, which completely depends on the tunable laser stability [20, 27].

Fig. 9 shows measured attenuation, GD and CD as function of wavelength for FBG with 55 GHz FWHM bandwidth. The insertion loss is 5,3 dB while its bandwidth at -1 dB level is 50 GHz and its bandwidth at -20 dB level is equal to 75 GHz. The group delay variation is limited to 50 ps in the pass-band and the dispersion at the center wavelength is equal to 0 ps/nm. The maximum dispersion in the bandwidth at -3 dB level is found to be within the range of -500 ps/nm to 500 ps/nm.

2.3. Efficient bandwidth measurement method

The main idea of our experiments is to evaluate efficient bandwidth of FBG with 55 GHz FWHM bandwidth. Efficient bandwidth of passive device provides limitations which are required to take into consideration for realization of DWDM-direct transmission system for broadband access.

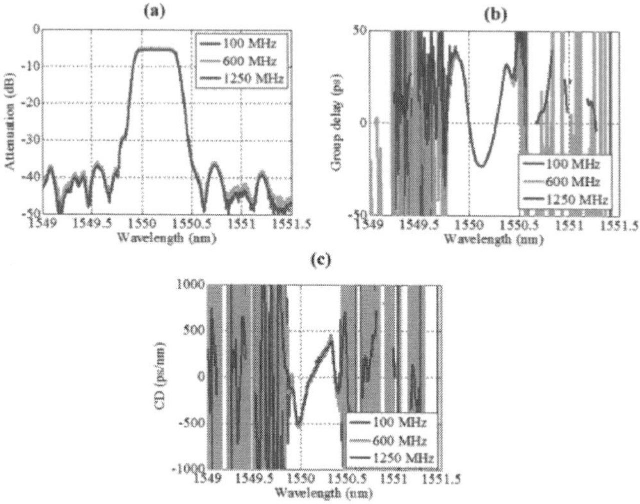

Figure 9. Measured attenuation (a), group delay (b), chromatic dispersion (c) and differential group delay (d) as function of wavelength for 55 GHz FBG filter

2.3.1. Method setup

Efficient bandwidth measurement scheme (see Fig. 10) was realized to investigate FBG with 55 GHz FWHM bandwidth with 2 GE and 10 GE optical signals. The efficient bandwidth measurement scheme consists of typical optical transmission system elements. The transmitter consists of pseudo-random data generator with 2^{31}-1 bit sequence, non-return to zero code former, continuous wavelength laser source and $LiNbO_3$-based external Mach Zehnder modulator.

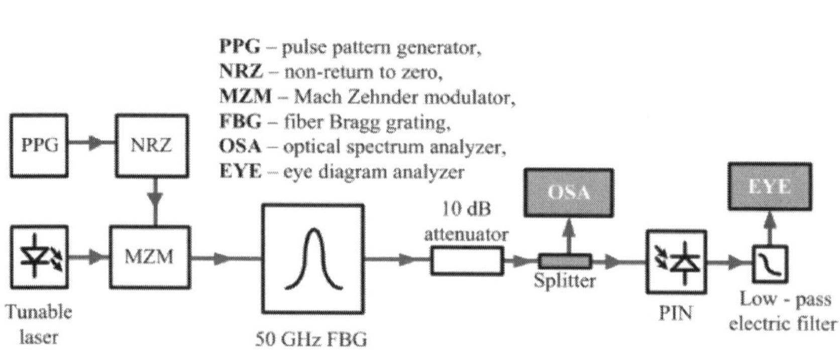

Figure 10. Efficient bandwidth measurement scheme

The data source produces a pseudo-random electrical signal, which represents the information. Then a code former is used to form NRZ code from incoming pseudo-random bit sequence. The optical pulses are obtained by modulating CW laser irradiation in MZM. Formed optical pulses are sent directly to a DUT at different CW laser central wavelength offset values. Receiver block consists of optical attenuator, PIN photodiode and Bessel – Thomson electrical filter (4 poles, 7.5 GHz -3dB bandwidth). Attenuator with 10 dB rated value was used to simulate loss of 20 km optical fiber, splicing and connectors in direct access systems. Oscilloscope and optical spectrum analyser (OSA) was used to perform measurements of eye diagram and optical power spectral densities, accordingly.

2.3.2. Results for 2.5 GE and 10 GE transmission speed

The bit error rate measurement is a simple method for systems performance evaluation. The error counting in a practical system for realistically low BER values (< 10^{-12}) can be a long process. Therefore the International Telecommunications union (ITU) has created the eye diagram masks for different bit rates with a definite BER value [28].

Fig. 11 shows the eye diagrams and optical power spectral densities of 2 GE optical signals after FBG with 55 GHz FWHM bandwidth for different laser central wavelength offset values (-0.2 nm, -0.1 nm, 0 nm, 0.1 nm, 0.2 nm). Offset value was changed within FBG device pass-band with 0.1 nm step. This value was chosen to fit DWDM systems wavelength grid defined in ITU-T G.694.1 recommendation. As we can see from results greater optical signal amplitude and phase distortions are at the edges of the band-pass optical filter. On Fig.11.a and Fig.11.e are shown eye diagrams for -0.2 nm and +0.2 nm offset values and there are signal waveform degradation. From these results FBG efficient bandwidth is 0.4 nm or 50 GHz and is the same as FWHM bandwidth.

Fig. 12 depicts out the eye diagrams and optical power spectral densities of 10 GE optical signals after FBG with 55 GHz FWHM bandwidth for different laser central wavelength offset values (same as in Fig. 11). On Fig.5.a and Fig.5.e are shown eye diagrams for -0.2 nm and +0.2 nm offset values and there are signal and mask crossing which means that defined BER value is exceeded. The results show that efficient bandwidth is 0.2 nm or 25 GHz and is 0.2 nm or 25 GHz lower than FWHM bandwidth for this transmission speed.

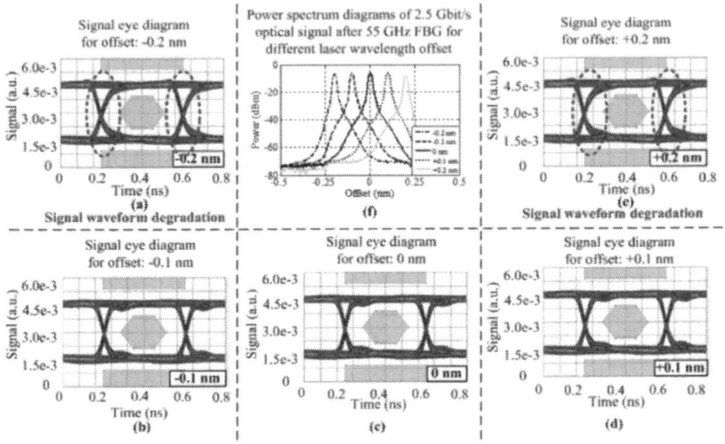

Figure 11. Eye diagrams (a-e) and optical power spectral densities (f) of 2GE optical signal after FBG with 55 GHz FWHM bandwidth for different CW laser wavelength offset shown in inset

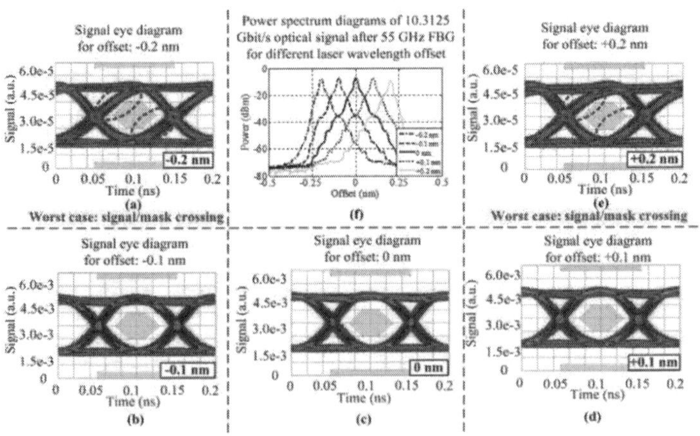

Figure 12. Eye diagrams (a-e) and optical power spectral densities (f) of 10GE optical signal after FBG with 55 GHz FWHM bandwidth for different CW laser wavelength offset shown in inset

3. REALIZATION OF NEW GENERATION ACCESS SYSTEM

Optical band-pass filter performance must be evaluated to increase spectral efficiency of overall optical transmission system [8]. Due to this, detailed investigation has been done into the FBG filter influence on the optical signals in DWDM-direct. For this purpose we have created the DWDM-direct measurement scheme and determined minimal channel interval for 55 GHz FBG filter at which the bit error ratio is sufficiently low. This evaluation was carried out employing eye diagrams and optical power spectral densities of the received optical signal.

3.1. Measurement setup

DWDM-direct scheme (see Fig. 13) is composed of three parts: a transmitter, an optical fiber, and a receiver. The transmitter consists of a pseudo-random data source with 2^{31}-1 bit sequence (Anritsu MU181020A), a non-return-to-zero code former (Anritsu MU181020A), a tunable continuous wavelength laser source (Agilent 81989A, 81949A), and an Avanex LiNbO3-based external MZM. The data source generates a pseudo-random electrical signal which contains the information to be transmitted via optical fiber. Then a code former is used to form an NRZ code from the incoming pseudo-random bit sequence. The optical pulses are obtained by modulating CW laser light in MZM with the generated bit sequence. After optical modulation the formed optical pulses are sent directly to a 20 km SSMF (G.652.d). The utilized fiber has a large core effective area of 80 μm^2, attenuation α = 0.2 dB/km, nonlinear refractive coefficient n_k = $2.5 \cdot 10^{-20}$ cm/W, and dispersion 16 ps/nm/km at the reference wavelength λ = 1550 nm. The receiver block consists of an optical filter (55 GHz FBG), a PIN photodiode, and a Bessel–Thomson's electrical filter (4 poles, 7.5 GHz -3dB bandwidth, Anritsu MP1026A) [16].

In this research the minimal channel interval for DWDM-direct systems with 55 GHz FBG filter has been determined employing measured eye diagrams an optical power spectrum densities. A high-frequency oscilloscope Anritsu MP1026A was used to perform the eye diagram measurements and the optical spectrum analyzer ADVANTEST Q8384 was employed to get optical power spectral densities.

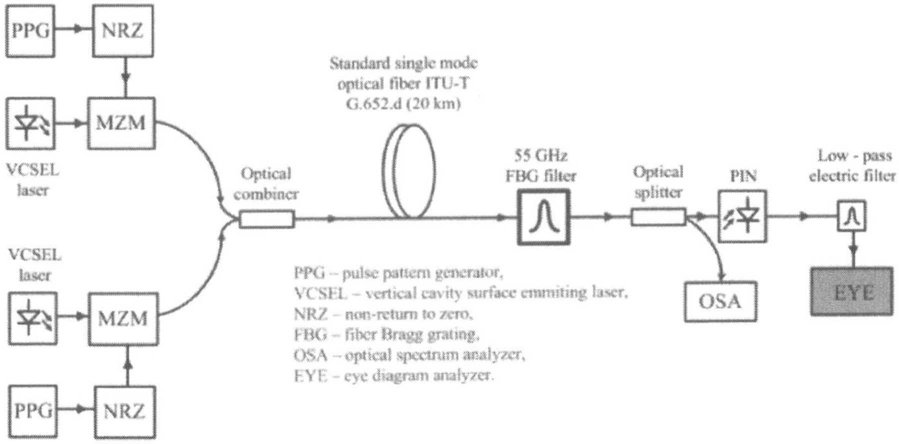

Figure 13. Realized DWDM– direct system for broadband access

3.2. Spectral efficiency enlargement

Fig. 14 shows the eye diagrams and optical power spectral densities of a 2.5 Gbit/s DWDM-direct system realized with 55 GHz FBG after 20 km of SSMF for different channel intervals from 25 GHz to 125 GHz with 25 GHz (0.2 nm in a wavelength range) step. The step value was chosen to fit DWDM wavelength grid defined in ITU-T G.694.1 recommendation. Both signal detection in the 2.5 Gbit/s DWDM system, 55 GHz FBG, was observed with a 25 GHz channel interval. To reduce undesirable adjacent signal, the channel interval was increased, which gave lower BER values for the detected signal. As a result, the adjacent channel was suppressed more efficiently, because the steepness of a 55 GHz FBG device is very good and the adjacent channel's isolation is ~35 dB. As one can see from the results (Fig. 14b), a 50 GHz channel interval is sufficient to ensure the appropriate BER value for adequate system's performance. The results for greater channel intervals (75 GHz, 100 GHz and 125 GHz, Fig. 14c–e) are also shown to demonstrate DWDM-direct system's stability in the employed spectral range.

The eye diagrams and optical power spectral densities of a 10 Gbit/s DWDM-direct system for broadband access after 20 km of SSMF for the same channel intervals as in the previous case are shown in Fig. 15. Due to a higher transmission speed, the optical power spectral density is broader, which results in stronger influence of CD on the signal quality. This leads to greater degradation of the optical signal, which emerges as a larger standard deviation and jitter for “0” and “1” levels in eye diagram. Similar to the above, a 50 GHz channel interval is sufficient to ensure the appropriate BER value for normal performance of the system at 10 Gbit/s transmission speed; the spectral efficiency is in this case improved from 0.18 bit/s/Hz to 0.2 bit/s/Hz.

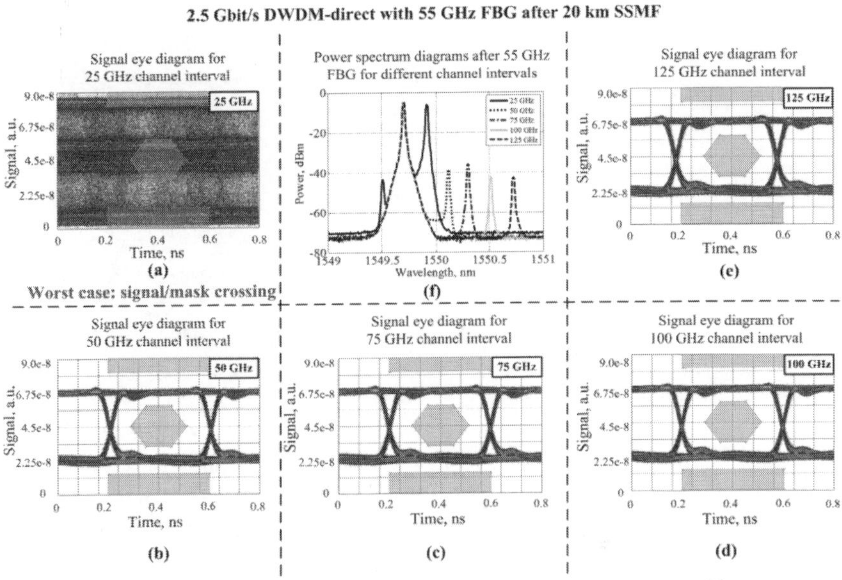

Figure 14. Eye diagrams (a–e) and optical power spectral densities (f) of 2.5 Gbit/s DWDM-direct system realized with a 55 GHz FBG after 20 km of SSMF for different channel intervals (shown in insets).

3.3. Channel number enlargement: numerical evaluation

Simulation scheme is shown in Fig. 3. Channel count of simulation scheme depends on simulation setup. Two, four and eight channels were chosen balancing between total capacity on one hand and physical limitations on the other.

The main idea of simulations is to demonstrate the possibility of channel number enlargement for FBG filter with 55 GHz FWHM bandwidth in DWDM-direct transmission system for broadband access.

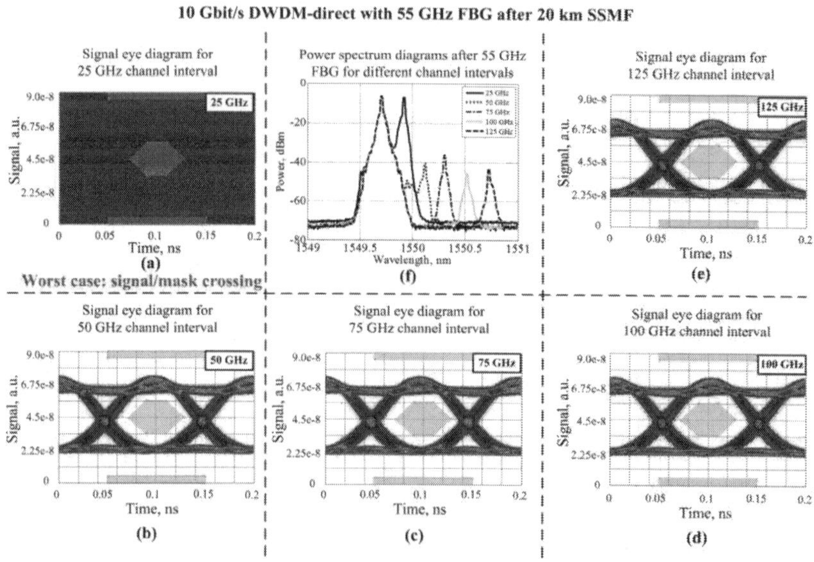

Figure 15. Eye diagrams (a–e) and optical power spectral densities (f) of 10 Gbit/s DWDM-direct system realized with a 55 GHz FBG after 20 km of SSMF for different channel intervals (shown in insets).

Fig.16.a-c. depicts out power spectral densities and eye diagrams of 2.5 Gbit/s DWDM-direct transmission system with different channel count after 20 km of SSMF and Fig.16.d. shows BER dependence on distance for 50 GHz FBG. We can see that adjacent channel isolation value (~ 30 dB) for 55 GHz FBG is sufficient to realize reliable transmission at eight channel case. In this case influence of adjacent channel caused impairments is minimized by proper optical band-pass filter parameter selection.

Fig.17.a-c. depicts out power spectral densities and eye diagrams of 10 Gbit/s DWDM-direct transmission system with different channel count after 10 km of SSMF and Fig.17.d. shows BER dependence on distance for 55 GHz FBG. Transmission at higher data speed is more affected by chromatic dispersion of optical fibre and total power budget of system is reduced because of greater excess loss in MZM and lower receiver sensitivity for appropriate BER threshold.

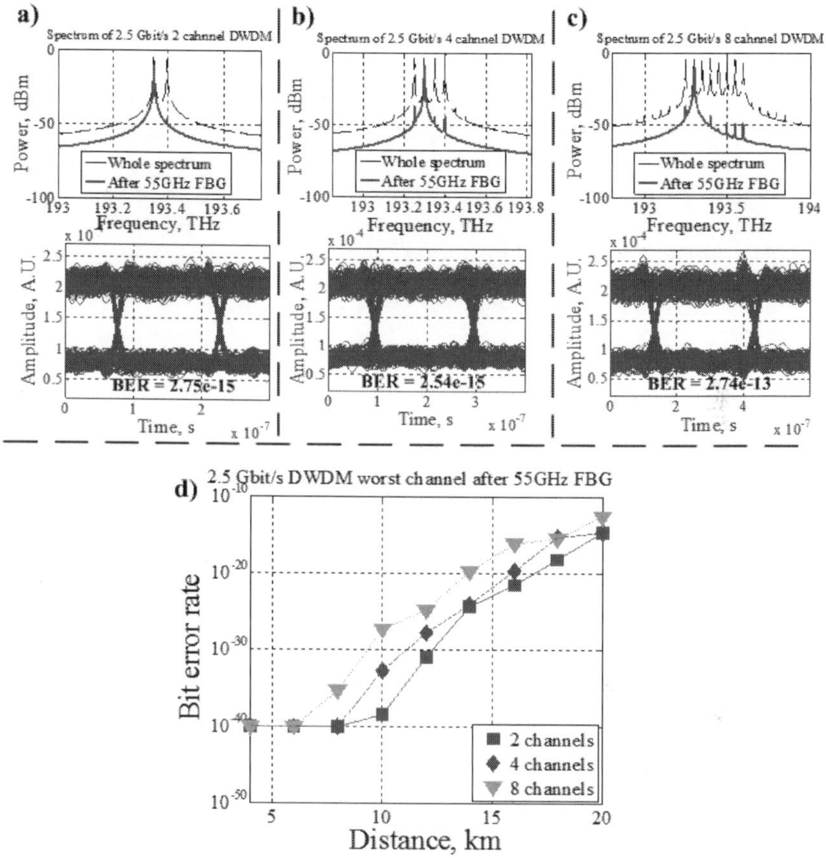

Figure 16. Power spectral densities and eye diagrams of 2.5 Gbit/s DWDM-direct a) two channels, b) four channels, c) eight channels system after 20 km of SSMF and d) BER vs. Distance with 55 GHz FBG optical filter. Results obtained at the worst channel.

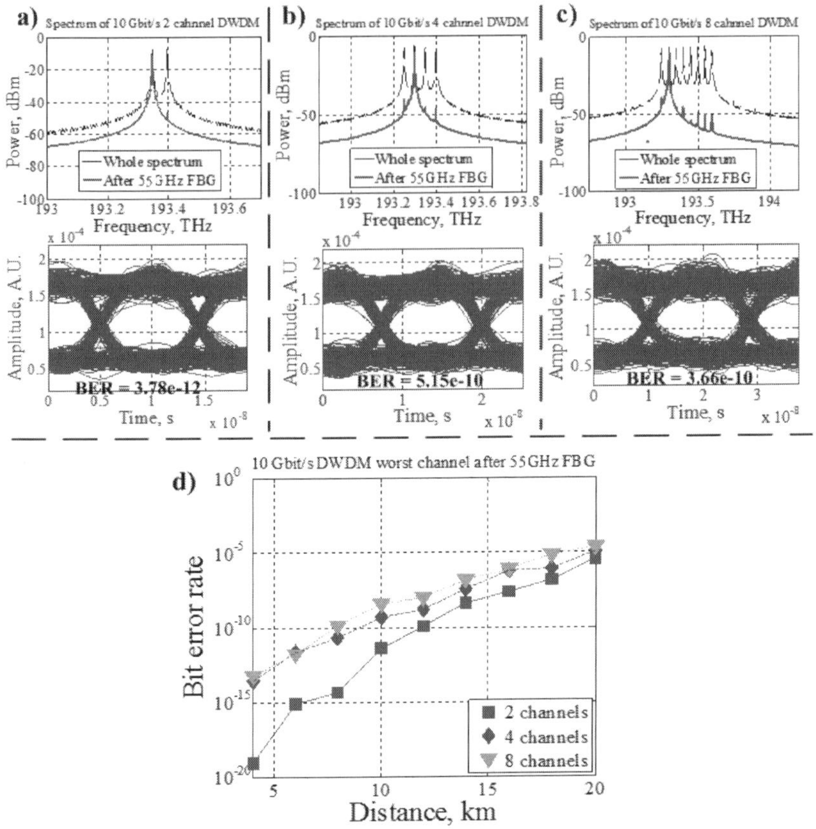

Figure 17. Power spectral densities and eye diagrams of 10 Gbit/s DWDM-direct a) two channels, b) four channels, c) eight channels system after 10 km of SSMF and d) BER vs. Distance with 55 GHz FBG optical filter. Results obtained at the worst channel.

4. CONCLUSIONS

As we can see from the results, the proper selection of optical filter amplitude transfer functions is of great importance. In this investigation influence of adjacent channel caused impairments is minimized by proper optical band-pass filter parameter selection. Reliable transmission is realized for 2.5 Gbit/s and 10 Gbit/s DWDM-direct with 55 GHz FBG for 20 km of SSMF.

Results show that DGD value within filters pass-band does not exceed 70 picoseconds for FBG with 50 GHz FWHM bandwidth. DGD value increases at the edges of the optical filter amplitude transfer function which could be a degradation factor for optical signal transmitted through device. Furthermore efficient bandwidth was evaluated for FBG devices employing optical signals with different transmission speed. Efficient bandwidth of 55 GHz FBG device was 0.4 nm or 50 GHz for 2GE optical signal and 0.2 nm or 25 GHz for 10GE optical signal.

We have realized a DWDM-direct system for broadband access that includes FBG filter with 55 GHz FWHM bandwidth. From the measurement results we found the minimal channel interval for the 55 GHz FBG to ensure reliable data transmission, and therefore were able to increase the spectral efficiency of the whole DWDM-direct system for broadband access. In 2.5 Gbit/s and 10 Gbit/s DWDM-direct systems with 55 GHz FBG the detection of both signals were observed for a 25 GHz channel interval. To achieve single-channel detection and suppression of the adjacent channel's power level the channel spacing was increased to 50 GHz. As a result the spectral efficiency of the 10 Gbit/s DWDM system with 55 GHz FBG was raised from 0.18 bit/s/Hz to 0.2 bit/s/Hz.

ACKNOWLEDGMENT

This work has been supported by the European Regional Development Fund in Latvia within the project Nr. 2010/0270/2DP/2.1.1.1.0/10/APIA/VIAA/002.

REFERENCES

1. L. G Kazovsky, W Shaw, T Gutierrez, D Cheng, N Wong, S. -W Next-generation, Optical Access Networks. Journal of Lightwave Technology 2007251134283442

2. E Wong, Next-Generation Broadband Access Networks and TechnologiesJournal Of Lightwave Technology20122012304597

3. J Hecht, Recycled Fiber Optics. How Old Ideas Drove New TechnologyOptics and Photonics News20122322229

4. N Kataoka, N Wada, W Xu, G Cincotti, and K Kitayama, I Gbps-class, bandwidth-symmetric, OCDM-PON system using hybrid multi-port and SSFBG en/decoder. 14th Conference on Optical Network Design and Modeling (ONDM) 2010March 15 2010.

5. F. J Effenberger, J Kani, and Y Maeda, Standardization Trends and Prospective Views on the Next Generation of Broadband Optical Access SystemsIEEE Journal on Selected Areas in Communications 2010286773780

6. E Kehayas, Designing Wavelength-Division-Multiplexed Optical Access Networks Using Reflective Photonic Components. 14th Conference on Optical Network Design and Modeling (ONDM) 2010March 15 2010.

7. G Agrawal, Nonlinear Fiber Optics (Third Edition). USA: Academic Press; 2001

8. J Segarra, V Sales, and J Prat, Agile Reconfigurable and Traffic Adapted All-Optical Access-Metro Networks" 11th International Conference on Transparent Optical Networks (ICTON) 2009June 28 2009.

9. O Ozolins, V Bobrovs, G Ivanovs, Efficient Bandwidth of 50 GHz Fiber Bragg Grating for New-Generation Optical Access. 19th Telecommunications forum TELFOR 2011November 2224

10. T Miyazawa, H Harai, Optical access architecture designs based on WDM-direct toward new generation networksIEICE Transactions on Communications2010E93B(2) 236-245.

11. M Pfennigbauer, ., P. J Winzer, . Choice of MUX/DEMUX filter characteristics for NRZ, RZ, and CSRZ DWDM systems. Journal of Lightwave Technology 2006; 24 4 16891696 .

12. Y Shen, K Lu, and W Gu, Coherent and Incoherent Crosstalk in WDM Optical Networks Journal Of Lightwave Technology1999175759764

13. H. S Phing, J Ali, R. A Rahman, and B. A Tahir, Fiber Bragg grating modeling, simulation and characteristics with different grating lengths. Journal of Fundamental Sciences 200731167175

14. R Kashyap, Fiber Bragg grating. USA: Academic Press; 1999

15. V Bobrovs, G Ivanovs, Parameter Evaluation of a Dense Optical Network. Electronics and Electrical Engineering 20064253337

16. O Ozolins, V Bobrovs, G Ivanovs, Efficient Wavelength Filters for DWDM SystemsLatvian Journal of Physics and Technical Sciences2010611324

17. H Venghaus, Wavelength filters in fibre opticsBerlin: Springer; 2006

18. M Azadeh, Fiber Optic engineering. London: Springer; 2009

19. L Chen, Z Zhang, X Bao, Polarization dependent loss vector measurement in a system with polarization mode dispersionOptical Fiber Technology20062006121251

20. R Hui, O Sullivan, M. Fiber optic measurement techniquesBurlington: Elsevier; 2009

21. J Haro, P. R Horche, Evolution of PMD with the temperature on installed fiber Optical Fiber Technology 2008143203213

22. X. S Yao, X Chen, T Liu, High accuracy polarization measurements using binary polarization rotatorsOptics Express201018766676685

23. X Kun, Y Dai, M Jin, F Jia, X Chen, X Li, S Xie, A novel method of automatic polarization measurement and its application to the higher-order PMD measurementOptics Communications200321512309314

24. Agilent TechnologiesAgilent N7788B/BD optical component analyser. Agilent Technologies, Inc. 2008116

25. M Westhauser, M Finkenbusch, C Remmersmann, S Pachnicke, P. M Krummrich, Optical Filter-Based Mitigation of Group Delay Ripple- and PMD- Related Penalties for High Capacity Metro Networks. Journal of Lightwave Technology2011291623502357

26. C Peucheret, Fibre and component induced limitations in high capacity optical networks. PhD thesis, Technical University of Denmark; 2004

27. Agilent TechnologiesAgilent 86038B Photonic Dispersion and Loss Analyzer, 2nd ed. Germany: Agilent Technologies Manufacturing GmbH; 2006

28. O Ozolins, V Bobrovs, G Ivanovs, DWDM Transmission Based on the Thin-Film Filter Technology.Latvian Journal of Physics and Technical Sciences201120113555

CHAPTER 9

Sensitivity-Improved Strain Sensor over a Large Range of Temperatures Using an Etched and Regenerated Fiber Bragg Grating

Yupeng Wang [1,*], Xueguang Qiao [1], Hangzhou Yang [1], Dan Su [1], Ling Li [1] and Tuan Guo [2,*]

[1] School of Physics, Northwest University, Xi'an, Shaanxi 710069, China

[2] Institute of Photonics Technology, Jinan University, Guangzhou 510632, China

ABSTRACT

A sensitivity-improved fiber-optic strain sensor using an etched and regenerated fiber Bragg grating (ER-FBG) suitable for a large range of temperature measurements has been proposed and experimentally demonstrated. The process of chemical etching (from 125 μm to 60 μm) provides regenerated gratings (at a temperature of 680 °C) with a stronger reflective intensity (from 43.7% to 69.8%), together with an improved and linear strain sensitivity (from 0.9 pm/με to 4.5 pm/με) over a large temperature range (from room temperature to 800 °C), making it a useful strain sensor for high temperature environments.

KEYWORDS

etched and regenerated fiber gratings; strain sensor; high temperature

1. INTRODUCTION

Diverse fiber Bragg gratings (FBGs) have been proposed and widely used as fiber-optic sensors to measure strain, displacement, refractive index and temperature due to their intrinsic advantages [1–3]. Among them, strain measurement must be performed in many high temperature environments, such as space plane engines, high temperature pipelines in the oil and gas industry, etc. To improve the temperature resistance of FBG for strain sensing, many techniques have been put forward in the past few years. Bragg gratings inscribed in Er-Sn-codoped germanosilicate fiber were employed to simultaneously measure the strain range of 0–1150 με at temperatures up to 500 °C with a strain sensitivity of 1.08 pm/με [4]. An alternative approach has been demonstrated by writing the grating in Sb-Er-Ge-codoped fiber, which enhances the temperature and strain sustainability to 600 °C and 2000 με, respectively, and a strain sensitivity of 1.15 pm/με is achieved [5]. Li et al. have reported the temperature response of chemical composite gratings over a temperature range from 24 °C to 900 °C and the strain response at a maximum applied strain of 1000 με, with a strain sensitivity of 1.1 pm/με at each temperature [6].

From the reports above, a strain sensor based on a bare FBG has a strain sensitivity of around 1 pm/με, which results in a low detection resolution. Conventionally, two approaches have been employed to enhance the strain sensitivity: one is recoating the fiber sensor with a layer of metal, the other is chemically etching the fiber sensor. A metal-packaged regenerated fiber Bragg grating (RFBG) using titanium–silver magnetron sputtering and nickel electroplating was developed for strain measurement [7], in which the strain sensitivity is 1.7 pm/με at temperatures ranging from room temperature to 400 °C. Lyons et al. presented a method involving chemically etched cladding to study the characteristic of FBGs and propose a tunable FBG filter, which has a repeated wavelength tuning range of 9 nm [8]. A FBG with a diameter of only 12 μm prepared through wet chemical etching has been employed to measure strain. The results show that it has ultra-high strain sensitivity up to 104.1 pm/με in the strain range only 0–100 με and temperature up to 65 °C [9].

Compared with the metal coating technique, chemical etching is a much easier and cost effective method to enhance the strain sensitivity of fiber sensor. Furthermore, RFBG has been used as an ideal candidate for operating at ultra-high temperatures [10–12]. In this work, we demonstrate an approach to enhance the reflective intensity of RFBG, improve its strain sensitivity and enlarge its temperature measurement range. The chemical etching has been performed on the seed gratings to get FBGs with different diameters. A standard thermal treatment technique is used to activate these seed gratings and obtain etched and regenerated fiber Bragg gratings (ER-FBGs). The results show the reflective intensity of ER-FBG can be effectively enhanced by reducing the fiber diameter. Meanwhile, an improved strain sensitivity (from 0.9 pm/με to 4.5 pm/με) over a large range of temperature measurements (from room temperature to 800 °C) have been experimentally achieved, making it a good candidate for strain sensors in ultrahigh temperature environments.

2. FABRICATION AND EXPERIMENT

The seed gratings (2.0 cm in length) are written in B/Ge codoped photosensitive fiber (PS1250/1500, Fibercore Ltd., Hampshire, UK). A continuous wave (CW) frequency-doubled argon ion laser with an operation wavelength of 244 nm and average output power of 28 mW is used. The grating is inscribed with well-established phase mask method. A phase mask with pitch period of 1072.6 nm is used for seed grating inscription. It should be noted that we use a laser beam scanning method (along fiber axial direction) to inscribe the grating. Therefore, the length of the grating is essentially dominated by the length of the phase mask (2.0 cm is used here).The fibers used (initial cladding diameter of 125 μm) are of 10 mol% of GeO_2 and 14–18 mol% of B_2O_3 and with core-cladding refractive index difference of 0.5%. Before grating inscription, the fibers are hydrogen loaded under a pressure of 11.3 MPa for 10 days at room temperature. The fabricated seed gratings are stored in a drying oven (50 °C) for two days to diffuse out the hydrogen. The grating strength then decays and stabilizes at 26 dB.

The etched FBGs with uniform fiber diameter are achieved by immersing them in HF acid solution (40% in water). During etching, the fiber diameter is measured by a microscope. As a result, the FBGs with different diameters of 125 μm, 90 μm and 60 μm are acquired, respectively. The fiber diameter of 60 μm is selected as a case, which increases the strain sensitivity without changing the FBG temperature and strain properties for large ranges of temperature and strain measurement. The experimental setup for thermal regeneration of FBG and the strain response at high temperature for etched RFBG is shown in Figure 1. Amplified spontaneous emission (ASE) from an erbium doped fiber amplifier (EDFA) is used as a broadband light source (BBS), which is connected through a circulator to the etched FBG. The etched FBG is positioned (stain free) in the tube furnace for isothermal annealing process. Both transmission and reflection spectra of the etched gratings are measured using an optical spectrum analyzer (OSA) with a wavelength resolution of 0.02 nm. Thermal regeneration is then applied for the seed gratings through a subsequent annealing process. To characterize the response of the regenerated gratings with longitudinally applied strain, the both sides of RFBG are fixed by using epoxy, in which one side is fastened on a micro-displacement stage with a displacement resolution of 0.01 mm. The fiber is then stretched by a calibrated micrometer and the measurements of both ΔL and L are taken. To make sure the grating provides a linear strain response for strain measurement, we pre-strained the grating before it is fixed (glued). The initial tension (pre-strain) can be measured by monitoring the wavelength shift of FBG before and after the grating is fixed, and the initial strain value is 0 uε. A displacement step of 0.1 mm applied to 1.0 m fiber is equal to a strain step of 100 με. Each increment and the corresponding λ_B are recorded, simultaneously.

Figure 2a shows the measured dynamics of grating regeneration for different diameter fibers. The color curves represent the grating regeneration dynamics of three grating samples with different diameters, while the black curve represents the temperature change at the corresponding time. The heating temperature is ramping at 10 °C/min to 680 °C at which the seed grating is totally immersed into the noise level. The fiber is then held at 680 °C for 30 min to activate the regeneration behavior of the grating. After that, the heating temperature is passively cooled down to room temperature. Figure 2b shows the reflection and transmission spectra of the seed gratings at room temperature (black curves) and their RFBGs with different diameters, respectively.

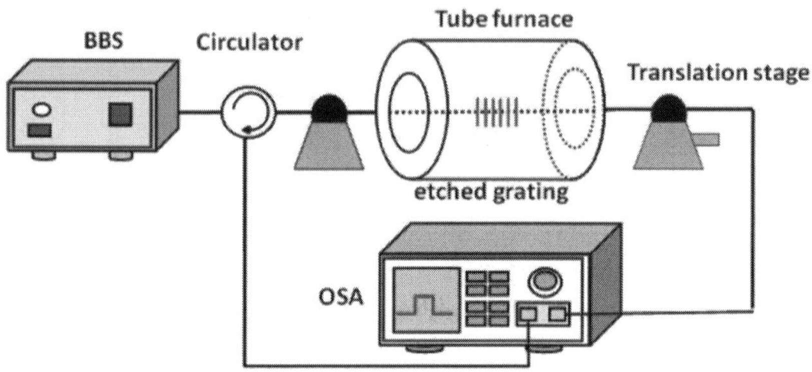

Figure 1. Schematic diagram of the experimental setup.

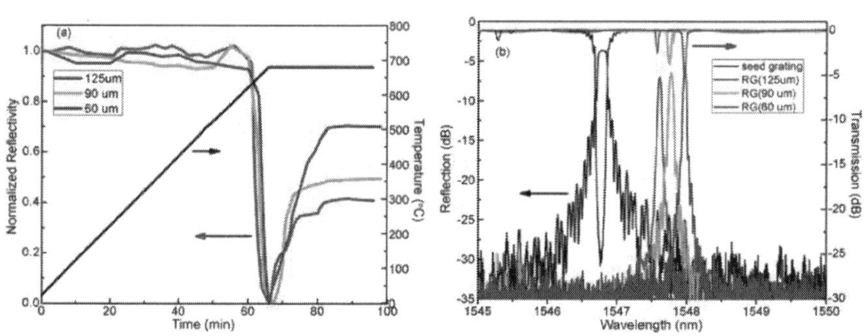

Figure 2. (**a**) Evolution of etched gratings reflectivity during thermal regeneration. (**b**) The reflection and transmission spectrum of seed (black line) and regenerated gratings, namely 125 μm (red curve), 90 μm (green curve) and 60 μm (blue curve) respectively.

3. RESULTS AND DISCUSSION

Clearly, according to Figure 2, we get reflectivity-improved regenerated FBGs via cladding etching, which correspond to reflectivity of 43.7%

(diameter of 125 μm), 55.3% (diameter of 90 μm) and 69.8% (diameter of 60 μm). This is because the condition of grating regeneration appears to be strongly correlated with differences in glass relaxation between the cladding and core [13]. The process of reducing the cladding diameter helps to accentuate this difference, and leads to stronger grating regeneration.

The temperature response of the RFBGs is tested from 25 °C to 800 °C. The temperature of the tube furnace is increased from room temperature to 800 °C by a controllable stepwise ramp of 10 °C/min. To avoid temperature fluctuations during the measurement process, the temperature of the tube furnace is kept constant for 20 min at each step of 100 °C to ensure uniform distribution before each recording. Figure 3 shows the temperature responses of three RFBGs with different diameters for increasing the temperature. The temperature sensitivities of 12.0 pm/°C, 12.1 pm/°C, and 12.2 pm/°C are obtained for fiber diameters of 125, 90 and 60 μm, respectively. The temperature sensitivities are almost constant with reducing the RFBG's diameter, which may be due to the fact that thermo-optic coefficient is consistent for that different diameters.

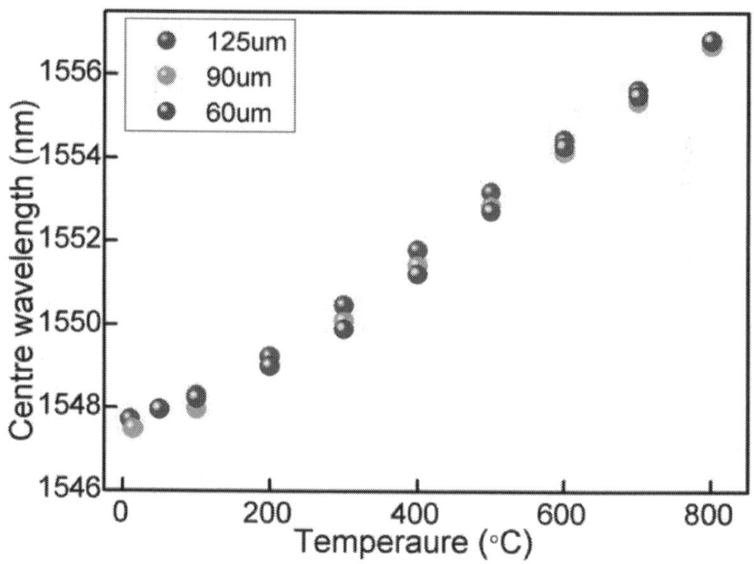

Figure 3. Wavelength shift of RFBGs with different diameters versus temperature range from 25 to 800 °C.

For a normal fiber grating, the applied strain will induce the shift of central wavelength due to co-effects of the change in grating period, and the shift in core index of refraction caused by the photo elastic effect. The strain effect of grating is given as:

$$\frac{\Delta \lambda_B}{\lambda_B} = (1 - p_e)\varepsilon \tag{1}$$

where ε is the applied strain in the axial fiber and p_e is the effective elastic-optic coefficient given by:

$$p_e = \frac{n_{eff}^{2}}{2}\left[p_{12} - \upsilon(p_{11} + p_{12})\right] \tag{2}$$

where p_{11} and p_{12} are the components of the silica strain-optic tensor, and υ is Poisson's ratio. For typical silica optical fibers, $p_{11} = 0.121$, $p_{12} = 0.270$, $\upsilon = 0.17$ and $n_{eff} = 1.456$, we can present a grating strain sensitivity coefficient K_ε, $K_\varepsilon = 1 - p_e = 0.784$. For a FBG with a center wavelength of 1550 nm, the expected strain sensitivity is approximately 1.2 pm/με. By reducing the cross sectional area of a fiber, etching can induce a significant reduction in the force required to produce a given strain [8]. It offers the possibility of improving the strain sensitivity for etched cladding FBG applied the same strain, compared with standard 125 μm FBG.

To test the strain response of proposed structure, strains are applied to RFBG at each constant temperature. The applied strains on the RFBG are increased from 0 to 1000 με then decreased back to 0 με by steps of 100 uε. At each step, the strain is held 10 min before reading the wavelength. The test results show that no evidence of hysteresis are observed since the highly repeatable and overlapped data are obtained for both increasing and decreasing strains. Figure 4 shows the strain response of RFBGs with different diameters for temperature from 100 °C to 800 °C, the results indicate that the wavelength shifts have good linearity for the applied tensile strain at each constant temperature. Figure 5a indicates that the strain response of RFBGs with different diameters at temperature of 700 °C, and the strain sensitivity is 0.9 pm/με for the RFBG with 125 μm, which is less than the numerical result of 1.2 pm/με of the normal FBGs. This is mainly due to that the different fiber used in our experiment is B/Ge codoped photosensitive fiber including 10 mol% of GeO_2 and 14–18 mol% of B_2O_3. On the other hand, the strain sensitivities of 1.7 pm/με and 4.5 pm/με for ER-FBGs with diameter of 90 μm and 60 μm respectively are obtained at 700 °C, and all the determination coefficients R^2 are 0.99. Compared with the RFBG with the diameter of 125 μm, the strain sensitivities with diameters of 90 μm and 60 μm are respectively increased around 2.0 and 5.0 times. By reducing the cladding diameter of a FBG through etching with HF, the strain sensitivity can be enhanced significantly. This is mainly due to the effect of stress concentration (geometric discontinuities cause an object to experience a local increase in the intensity of a stress field). The etched fiber section may suffer more stress than that of standard fiber both sides, leading to improved strain sensitivity.

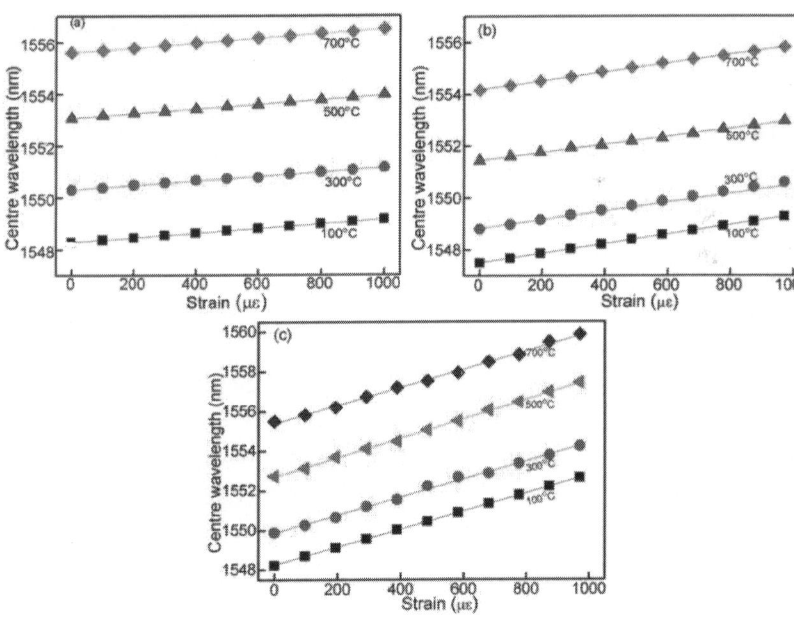

Figure 4. Wavelength shift of ER-FBGs versus axial strain at various temperatures: RFBG diameters of (**a**) 125 μm, (**b**) 90 μm, and (**c**) 60 μm.

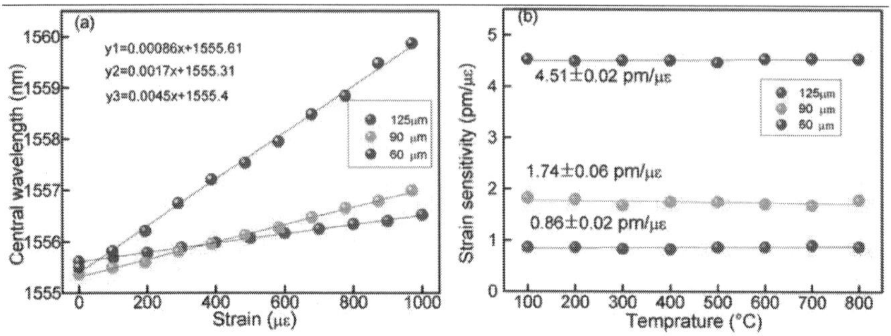

Figure 5. Strain-sensitivity of ER-FBGs with different cladding diameters (**a**) and their stable strain-sensitivity over the temperature from 100 °C to 800 °C (**b**).

The strain sensitivities are obtained by repeating the above-mentioned experiment many times. The results are shown in Figure 5b, in which the strain sensitivities of ER-FBGs with different diameters at various setting temperatures are 0.9, 1.7 and 4.5 pm/με, corresponding to 125, 90 and 60 μm RFBGs respectively. The results in Figure 5 denote that the strain sensitivity remains constant at each temperature setting. It should be noted that the ER-FBGs operated at high temperature do not reach the glass-softening region and still maintain a reproducible stress response.

4. CONCLUSIONS

Axial strain responses of ER-FBGs with different diameters have been studied over a large temperature range. The process of chemical etching (from 125 μm to 60 μm) enables the regenerated fiber gratings (at temperature of 680 °C) with a stronger reflective intensity (from 43.7% to 69.8%) together with an improved and linear strain sensitivity (from 0.9 pm/με to 4.5 pm/με) over a large temperature measurement range (from room temperature to 800 °C), making it a good strain sensor candidate for high temperature environments like aerospace engines, and high temperature pipelines in the oil and gas industry.

ACKNOWLEDGMENTS

This work was supported by the National Natural Science Foundation of China (Nos. 60727004, 61077060, 61205080, 61235005, 61327012, and 61405160), National High Technology Research and Development Program 863(Nos. 2007AA03Z413, 2009AA06Z203), Ministry of Education Project of Science and Technology Innovation (No. Z08119), Ministry of Science and Technology Project of International Cooperation (No. 2008CR1063), Shaanxi Province Project of Science and Technology Innovation (Nos. 2009ZKC01-19, 2008ZDGC-14).

AUTHOR CONTRIBUTIONS

In this article, Yupeng Wang and Hangzhou Yang designed the experiment and wrote the manuscript. Dan Su and Ling Li collected experimental data. After the manuscript was finished, Xueguang Qiao, Tuan Guo took charge of modifying the article.

CONFLICTS OF INTEREST

The authors declare no conflict of interest.

REFERENCES

1. Bai, Z.Y.; Zhang, W.G. Simultaneous measurement of strain and temperature using a long period fiber grating based on waist-enlarged fusion bitapers. J. Opt. **2014**, 16.

2. Rong, Q.Z.; Qiao, X.G.; Zhang, J.; Wang, R.H.; Hu, M.L; Feng, Z.Y. Simultaneous measurement for displacement and temperature using fiber Bragg grating cladding mode based on core diameter mismatch. J. Light. Technol. **2012**, 30, 1645–1650.

3. Ma, Y.; Qiao, X.G.; Rong, Q.Z.; Wang, R.H.; Zhang, J.; Hu, M.L.; Feng, Z.Y. Reflective fiber-optic refractometer based on a thin-corefiber tailored Bragg grating reflection. Opt. Lett. **2012**, 37, 323–325.

4. Pal, S.; Sun, T.; Grattan, K.V.T.; Wade, S.A.; Collins, S.F.; Baxter, G.W.; Monnom, G. Bragg grating performance in Er-Sn-doped germanosilicate fiber for simultaneous measurement of wide range temperature (to 500 °C) and strain. Rev. Sci. Instrum. **2003**, 74, 4858–4862.

5. Pal, S.; Shen, Y.; Mandal, J.; Sun, T.; Grattan, K.T. Simultaneous measurement of strain (to 2000 με) and temperature (to 600 °C) using a combined Sb-Er-Ge-codoped fiber-fluorescence and grating-based technique. IEEE Sens. J. **2005**, 5, 1462–1468.

6. Li, G.Y.; Guan, B.O. The strain response of chemical composition gratings at high temperatures. Meas. Sci. Technol. **2009**, 20.

7. Tu, Y.; Tu, S.T. Fabrication and characterization of a metal-packaged regenerated fiber Bragg grating strain sensor for structural integrity monitoring of high-temperature components. Smart Mater. Struct. **2014**, 23.

8. Lyons, E.R.; Lee, H.P. Demonstration of an etched cladding fiber Bragg grating filter with reduced tuning force requirement. IEEE Photon. Tech. Lett. **1999**, 11, 1626–1630.

9. Ferreira, M.S.; Bierlich, J.; Becke, M.; Schuster, K.; Santos, J.L.; Frazão, O. Ultra-high sensitive strain sensor based on post-processed optical fiber Bragg grating. Fibers **2014**, 2, 142–149.

10. Bandyopadhyay, S.; Canning, J. Ultrahigh-temperature regenerated gratings in boron-codoped germanosilicate optical fiber using 193 nm. Opt. Lett. **2008**, 33, 1917–1919.

11. Yang, H.Z.; Chong, W.Y.; Cheong, Y.K.; Lim, K.S.; Pua, C.H.; Harun, S.W.; Ahmad, H. Thermal regeneration in etched-core fiber Bragg grating. IEEE Sens. J. **2013**, 13, 2581–2585.

12. Yang, H.Z.; Chong, W.Y.; Qiao, X.G.; Lim, M.J.; Lim, K.S.; Islam, M.R.; Ali, N.M.; Ahmad, H. 1.3 μm and 1.55 μm thermally regenerated gratings in hydrogenated Boron/Germanium co-doped photosensitivity fiber. IEEE Sens. J. **2014**, 14, 1352–1356.

13. Wang, T.; Shao, L.Y.; Canning, J.; Cook, K. Temperature and strain characterization of regenerated gratings. Opt. Lett. **2012**, 38, 247–249.

CHAPTER 10

Long-Period Fiber Gratings in Active Fibers

David Krčmařík[1], Mykola Kulishov[2] and Radan Slavík[1, 3]

[1] Institute of Photonics and Electronics AV CR, v.v.i, Praha, Czech Republic

[2] HTA Photomask, 1605 Remuda Lane San Jose, CA, U.S.A.

[3] Optoelectronics Research Centre, University of Southampton, Southampton, Great Britain

1. INTRODUCTION

Traditionally, long period fiber gratings (LPG) are made in passive optical fibers that have negligible loss. However, loss or gain that can be controlled via optical pumping adds a new degree of freedom and – as will be shown in this chapter – brings many new and interesting properties.

From the historical perspective, the first attempt to combine the fiber gain and LPG filtering characteristics was for gain flattening of an Erbium-doped fiber (EDF) amplifier by inscribing LPG directly into the active fiber [1]. At the same time, theoretical studies [2,3] showed that a proper level of loss/gain in the fiber core or cladding can modify the LPG transmission characteristics. Significant theoretical and experimental body of work has been published since with new emerging applications appearing.

In this chapter, we investigate the new phenomena brought by the presence of the loss/gain [2,3]. Following this, we look on practical possibilities how to obtain required gain in active optical fibers and show how to analyze such structures, in which (incoherent) noise from an amplifying fiber is simultaneously generated and diffracted at the LPG [4]. Finally, we discuss possible application of the LPG in active fibers.

2. THEORETICAL ANALYSIS

First, we analyze LPG using standard coupled mode theory in which we consider that the refractive index is a complex number in which the imaginary part represents the gain/loss. Following this analysis, we show approaches into rigorous modeling of the active gain medium that contains an LPG, including the spontaneous emission and amplified spontaneous emission generation, propagation, and diffraction.

2.1. Coupled mode equations considering gain/loss

Similarly to an LPG in a gain/loss-less fiber, LPG assists coupling between the core and a cladding mode at the wavelength for which the phase matching condition is satisfied for real part of the mode propagation constants:

$$\mathrm{Re}\left(\beta^{cl}\right) = \mathrm{Re}(\beta^{co}) - \frac{2\pi}{\Lambda}, \tag{1}$$

where $\beta^{cl} = (2\pi/\lambda)n^{cl}_{eff}+j\alpha^{cl}$ and $\beta^{co} = (2\pi/\lambda)n^{co}_{eff} + j\alpha^{co}$ are the propagation constants of the interacting cladding and core modes, respectively, λ is the wavelength, Λ is the grating period, n^{cl}_{eff} and n^{co}_{eff} are the effective refractive indices of the interacting cladding mode and the core mode, α^{cl} and α^{co} are the absorption (when positive) or amplification (when negative) factors of the cladding and the core modes, respectively. Propagation constants for the core and cladding modes can be straightforwardly calculated, especially when considering a step-index fiber refractive index profile [4-6]. The strength of the coupling between the two modes depends on the coupling coefficient calculated as a confinement factor between the field of the two interacting modes [7]:

$$\kappa(z) = 0.5\omega\varepsilon_0 n^2 \sigma(z) \int_0^{2\pi} d\phi \int_0^{r_1} [r(E_r^{cl} E_r^{co^*} + E_\varphi^{cl} E_\varphi^{co^*})] dr \tag{2}$$

where ω is the angular frequency, ε_0 is the permittivity of vacuum, $\sigma(z)$ is the LPG index modulation amplitude, E_r and E_φ are the radial and azimuthal parts of electric field. The asterisk signifies complex conjugate value. For non-uniform LPGs the index modulation amplitide generally varies along the fiber, so it depends on the distance z from the LPG input. In the following text we consider only LPGs which are uniform ($\sigma(z)=const$ over the entire LPG length).

The mode coupling process along the LPG is described by the following coupled mode equations [7]:

$$\frac{dA}{dz} = i\kappa B \exp(-2i\delta z), \tag{3}$$

$$\frac{dB}{dz} = i\kappa A \exp(2i\delta z), \tag{4}$$

$$\delta = \frac{\beta^{co} - \beta^{cl}}{2} - \frac{2\pi}{\Lambda}, \tag{5}$$

where $A(z)$ and $B(z)$ are slowly varying envelopes of the core mode and the interacting cladding mode, and z is the coordinate originated at the LPG input. The solution of these equations for a uniform LPG of length L and for the initial condition of input signal launched in the core mode only ($A(0) = 1$ and $B(z) = 0$) can be written in the form of the amplitude core-to-core mode transmission [2]:

$$t = \frac{A(L)}{A(0)} = \left[\cos(\gamma L) + j\frac{\delta}{\gamma}\sin(\gamma L)\right]\exp\left[jL\left(\beta^{co} - \delta\right)\right], \tag{6}$$

where γ is:

$$\gamma = \sqrt{\delta^2 + |\kappa|^2}. \tag{7}$$

2.2. Condition for critical coupling

In a standard fiber like SMF-28, propagation losses are low and can be completely neglected over few centimeters, which is a typical LPG length, even for high order cladding modes (provided the protective high-refractive index fiber jacket is stripped off). Therefore, the long period index perturbation along the fiber length provides constant optical power exchange between the coupled modes. An optical signal launched into the core at the LPG resonance wavelength will be completely coupled by the LPG into the cladding mode and then it will be completely coupled back into the core. This process will be repeated again and again as long as long-period perturbation is induced along the fiber. At the LPG resonance wavelength where the mismatch factor $\delta=0$, the core-to-core mode transmission is

$$t = \cos(\kappa L). \tag{8}$$

The only way to provide full signal out-coupling from the core into the cladding mode is to terminate the LPG at discrete length values of $L = m\pi/(2\kappa)$, for $m = 1, 2, 3, \ldots$

As we will describe now, the performance changes completely when there is an appropriate level of loss or gain in the fiber core or cladding. For example, for the case of cladding mode loss, optical signal is getting lost there and less optical power returns back into the core. However, the situation is more complicated/interesting as shown below.

At the wavelength where the mode matching condition (Eq.(1)) is satisfied, the mismatch factor becomes purely imaginary: $\delta=-j(\alpha_{cl} - \alpha_{co})/2$. Substituting this expression for mismatch factor at the resonance wavelength into Eq.(6), and equating it to zero, we get a relationship between the grating strength, κL, and the dimensionless loss factor, $q= (\alpha_{cl} - \alpha_{co})/(2\kappa)$:

$$t = \cos(\sqrt{1-q^2}\,\kappa L) + \frac{q}{\sqrt{1-q^2}}\sin(\sqrt{1-q^2}\,\kappa L) = 0 \tag{9}$$

The solution of Eq.(9) is presented (in dimensionless variables: κL and q) in Fig. 1. For each value of κL the curve gives us the optimum ratio between the attenuation and the coupling coefficient (characterized by q) that provides full signal light attenuation in the core mode. Thus, solution exists for any κL, unlike for a loss-less LPG, where full coupling is possible only at discrete values of κL. In other words, zero core mode transmission for an LPG of any strength κL can be found, depending on the loss/gain relations of the core and the interacting cladding modes.

It is worth analyzing which parts of the curve shown in Fig. 1 are practically attainable. For example, in a usual situation in which the core mode loss is smaller than the cladding mode loss, we get always $q>0$ and thus full coupling is possible (from Fig. 1) only for $\kappa L>\pi/2$. However, $q<0$ (and thus also full coupling at $\kappa L<\pi/2$, Fig. 1) can be, e.g., achieved by introducing loss into the core mode (that is larger than the cladding mode loss). Another interesting possibilities are considering core mode gain ($q>0$ and thus solutions limited to $\kappa L>\pi/2$) or cladding mode gain with the core mode unamplified ($q<0$ with solutions for $\kappa L<\pi/2$) [3]. The latter could be practically achieved when considering double-clad fibers used for high-power fiber lasers, in which the cladding is doped by active ions [8].

Another interesting feature when considering gain/loss of core/cladding modes is that it may reduce interference sidelobes occurring outside of the resonance, obtaining a smooth, side-lobe free transmission spectrum [2].

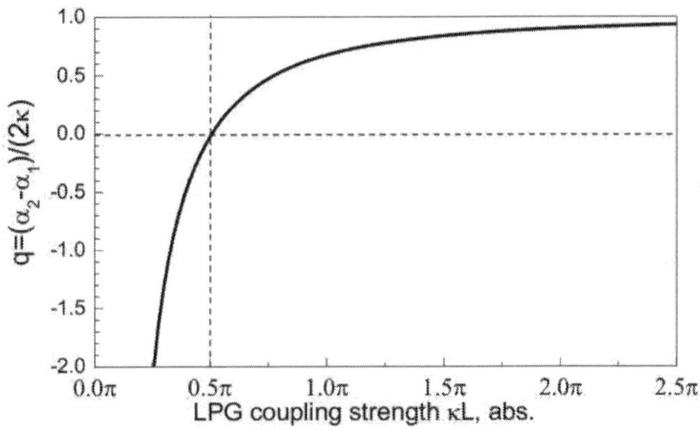

Figure 1. Required value of $q=(\alpha_{cl} -\alpha_{co})/(2\kappa))$ for obtaining full attenuation in the transmission spectrum of the core mode at the resonance wavelength (critical coupling) for a given LPG strength of κL.

To provide some physical insight into the mode coupling process in the presence of gain or loss, we show the grating transmission as a function of κL for various values of $q=\alpha_2/2\kappa$ (for clarity, only loss/gain in the cladding mode is considered here) – Fig. 2 [3]. The conventional sinusoidal transmission of loss/gain-free grating is shown as a solid curve. We can clearly see distinct types of transmission grating behavior depending on the magnitude of loss/gain.

2.3. Active fibers

The analysis of LPG in active fibers is based on a combination of two processes: coupled mode interaction (characterized by Eqs.(3-4)) and mode amplification (characterized by rate equations) [4]. To observe effects described in the previous section, high concentration of Er ions is required to achieve a significant gain over a length of an LPG that is usually relatively short (cm to tens of cm). Such high Er concentrations generally lead to cluster formations and up-conversion in the EDF, which has also to be taken into account.

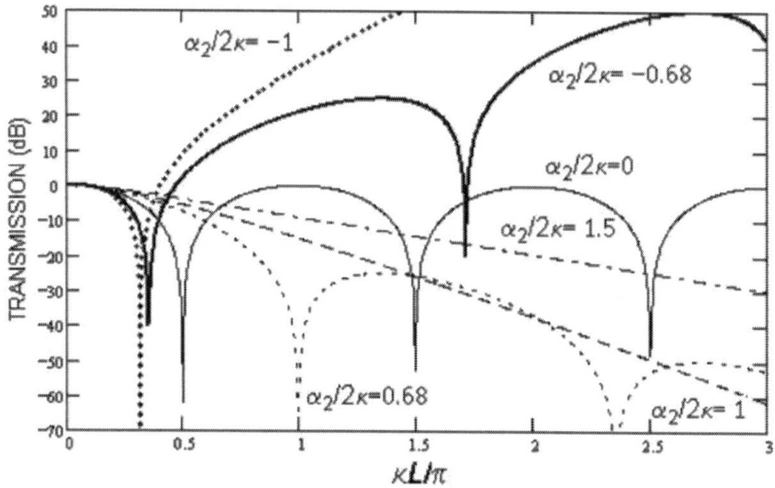

Figure 2. Variation of the grating transmission at the resonance wavelength with the grating strength κL for different $q=\alpha_2/2\kappa$ values.

There are several new phenomena occurring when considering not only the gain factor as in the previous part. The presence of gain/loss in the fiber with an LPG also changes resonance wavelength through modification of the effective refractive index of the coupled modes through Kramers-Kroning effect. Further, the presence of (amplified) spontaneous emission, (A)SE limits the maximum attainable transmission dip [4].

Here, the fibre is simulated by a two level system with the ground level and the meta-stable (excited) level. Due to high concentration of *Er* ions in the fibre it is necessary to take into account possible cluster formation and elevated occurrence of the up-conversion (process when the *Er* ions reach higher levels of excitation). Both cluster formation and up-conversion severely limits the maximum attainable gain for fibres with high concentration of Er ions and thus has to be considered.

In the up-conversion model we assume that the probability of the interaction of an ion with its neighbour is proportional to N_2^2, where N_2 is the number of Er ions in the excited state. Next model assumes cluster formation. In a cluster, only a single Er ion can contribute to the gain. In our analysis, clusters with more than two Er ions are not considered, as their contribution is generally negligible. In a combined model [9] the total number of ions contributing to the gain is:

$$N_2 = N_2^{SC} + N_2^{P}, \tag{10}$$

where N_2^{SC} is an average number of single excited ions and N_2^{P} is an average number of excited clusters. The average number of Er ions in not-excited state is $N_1 = N_{Er} - N_2$ (with N_{Er} total number of Er ions). Average number of single excited ions and excited clusters is:

$$N_2^{SC} = \frac{(1-2R)N_{Er}\sum_k \frac{P_k(z)\sigma_k^a \tau_{21}\Gamma_k}{h\nu_k A_{eff}}}{1 + C_{up}(1-2R)N_2^{SC}\tau_{21} + \sum_k \frac{P_k(z)}{P_k^{IS}}}, \tag{11}$$

$$N_2^{P} = \frac{2RN_{Er}\sum_k \frac{P_k(z)\sigma_k^a \tau_{21}\Gamma_k}{h\nu_k A_{eff}}}{1 + \sum_k \frac{P_k(z)}{P_k^{IS}} + \sum_k \frac{P_k(z)\sigma_k^a \tau_{21}\Gamma_k}{h\nu_k A_{eff}}}, \tag{12}$$

Where

$$P_k^{IS} = \frac{h\nu_k A_{eff}}{(\sigma_k^a + \sigma_k^e)\tau_{21}\Gamma_k}.$$

(13)

In the above equations R is a percentage of ions in clusters, P_k is the peak power of signals and pump on ν_k frequencies, τ_{21} is the metastable state lifetime, h is the Planck constant, A_{eff} is the effective core area and σ_k^a with σ_k^e are the absorption and emission cross-sections defined in the characteristics of the fiber. C_{up} [10] is the concentration independent and host dependent constant in m^3/s. Constant C_{up} can be determined by fitting the measured data. Γ_k is the overlap integral between the doped area and the optical mode field [11].

In an active medium we can observe a refractive index change due to the pump power. This translates into the slight wavelength shift of the coupling wavelength. The refractive index change is given by [12]:

$$\delta n = \left(\frac{\sigma_e' N_{Er}\alpha}{1+\alpha} - \frac{\sigma_a' N_{Er}}{1+\alpha} \right) \frac{\lambda \Gamma(\lambda)}{4\pi}, \tag{14}$$

$$\alpha = \frac{\sum_k P_k(z)}{P_{thr}}, \tag{15}$$

where P_{thr} is the threshold power for which the number of Er ions in the ground state is the same as in the meta-stable state ($N_1=N_2$). P_k sums both the signal powers and the pump powers in forward and backward directions. Since the pump propagates mainly in the core, we have neglected any refractive index changes in the cladding. Real parts of the absorption coefficient σ_a' and the emission coefficient σ_e' are computed from σ_a and σ_e through Kramers-Kroning relations.

To include the active medium into the coupled-mode theory, it is necessary to define the signal absorption coefficient $g_a(z, \nu)$, signal emission coefficient $g_e(z, \nu)$ and pump absorption coefficient α_p [13]:

$$g_a(z,\nu) = \sigma^a(\nu)N_1(z)\Gamma(\nu) - BL, \tag{16}$$

$$g_e(z,\nu) = \sigma^e(\nu)N_2(z)\Gamma(\nu), \tag{17}$$

$$\alpha_p(z,\nu_p) = \sigma_p^a(\nu_p)N_1(z)\Gamma_p(\nu_p) + BL, \tag{18}$$

where the index p is attributed to the variables corresponding to the pump, and BL is the background loss. Here we assume a fiber with Er ions doped in the core only and thus that only the core mode is amplified since most of the power of the cladding modes propagates in the cladding. The core mode is amplified according to:

$$\frac{dP_A}{dz} = (g_e - g_a)P_A, \tag{19}$$

where P_A is the core mode intensity. The pump is then described similarly by:

$$\frac{dP_p^{\pm}}{dz} = -\alpha_p P_p^{\pm}. \tag{20}$$

The signs ± mean forward and backward pumping.

2.4. Amplified spontaneous emission and modes of pumping

As described above, more careful approach has to be taken for ASE. The coupled mode equations, Eqs.(3-4), are amplitude and phase dependent (describing coherent interaction), while the rate equations represented by Eqs.(16-20) are power intensity-dependent only (describing an incoherent interaction).

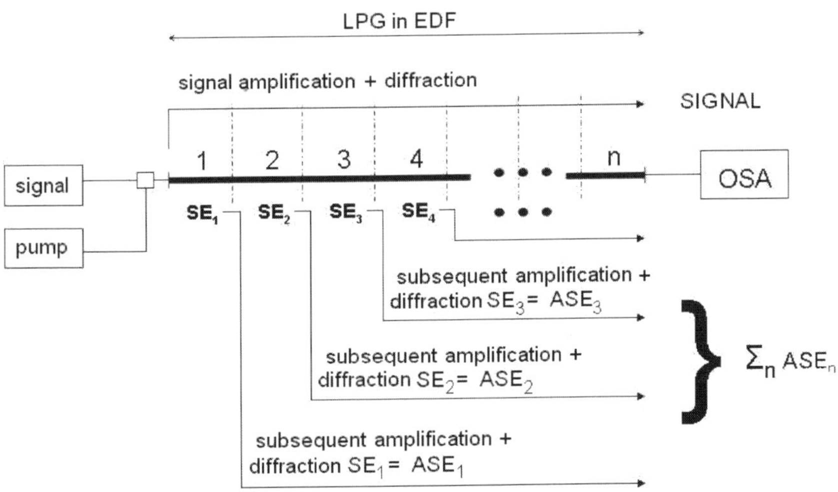

Figure 3. Description of SE and ASE treatment (SE – spontaneous emission, ASE – amplified spontaneous emission, OSA – optical spectrum analyzer).

ASE is seeded along the entire EDF length by spontaneous emission (SE) and subsequently amplified. In a standard EDF analysis, the fiber is divided into *n* segments first, the considered spectral bandwidth is divided into slots and each spectral component is propagated separately. Subsequently, the noise present at the input of the i-th segment, i=1, 2,..n (SE generated in the (i-1)-th segment and ASE amplified in the (i-1)-th segment) is amplified (forming ASE at the output of the i-th segment) and summed with SE generated in the i-th segment. Thus, at the output, the ASE is represented by a single number (for each spectral component). Here, we divide the fiber in the same manner as described above. However, we have to prevent summing of SE that was generated at different positions along the LPG: the SE generated at the beginning of the fiber undergoes diffraction along the entire length of the LPG, SE generated in its middle is diffracted at one half of the LPG, while the SE generated at its end is not diffracted at all. Thus, we take SE generated in the first segment and subsequently amplify it and diffract along the full length of the LPG. We continue similarly with the SE generated in the 2nd, 3rd … nth segments. As a result, we have n contributions of the SE+ASE at the output (each coming from SE generated in the i-th segment, $i=1,2..n$ and subsequently amplified in the n-i segments) at the output. These contributions are incoherent (as each of them contains photons generated by SE) and thus can be summed in power intensity at the output, representing the ASE total power (for each spectral component). This approach is graphically shown in Fig. 3.

In practice, we generate SE and propagate ASE within each segment according to the equation:

$$\frac{dP_{ASE}^{\pm}}{dz} = \pm 2h\nu(\Delta\nu)g_e \pm (g_e - g_a)P_{ASE}^{\pm}, \tag{21}$$

where Δv is the frequency slot. At the end of the segment the resulting ASE gets diffracted and new SE starts to arise at the beginning of the next segment.

Finally, we implemented also analysis for the contra-directional pumping configuration. In this scheme, the pump and the signal are propagated in opposite directions. For this configuration, the simulated data were obtained iteratively [4].

An example of the results using the theoretical analysis (populations of the meta-stable levels N_2 and ground levels N_1 of the EDF together with the pump power) is shown in Fig. 4. Here, we used parameters of the EDF later used in experiments – it is Liekki Er80-8/125 – more details can be found in Section 3.

Detailed analysis presented in [4] Fig. 5) reveals that optimal pumping scheme has to be chosen according to a specific application. Backward pumping is better when we want to limit the nonlinearities caused by an excessive power at the beginning of the fiber. On the other hand forward pumping shows better results for ASE suppression. For parameters considered, the LPG was always pumped relatively uniformly along its length. Therefore the transmission at the resonance wavelength is expected to be limited by the ASE formation rather than the non-uniform LPG strength that would be due to non-uniform pumping along the grating.

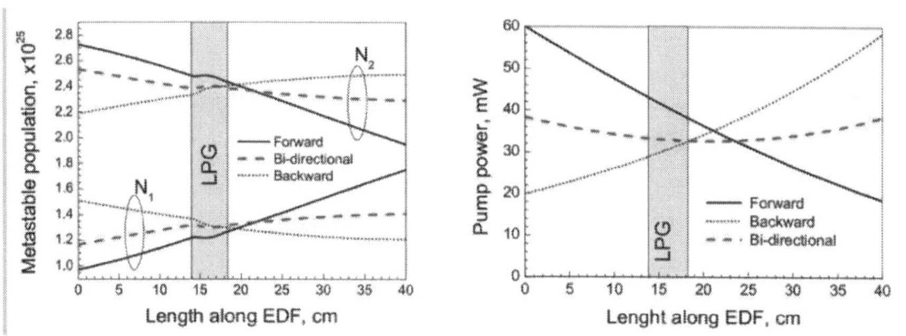

Figure 4. N_1 and N_2 populations (left) and pump power (right) along 40-cm long EDF sample with a 5-cm long LPG inscribed [position shown in the graphs] for total pump power of 60 mW and the input signal power of 100 μW.

3. CHOICE OF EDF AND FABRICATION TECHNIQUES

Most of commercially-available EDFs have relatively high numerical aperture (0.2-0.4), which helps in obtaining good performance of these fibers. This, however, requires LPGs with sub-100-μm period, as the effective refractive index of the core mode is significantly higher than in low-numerical aperture fibers (e.g., telecom SMF-28 with numerical aperture of 0.12). Unfortunately, this is prohibitive for CO_2 [14] and arc [15] LPG inscription techniques – both of them operate on a principle of a local heating of the fiber and it is difficult to heat a fiber with 125 μm diameter over length significantly shorter than that. The techniques suitable are UV-writing [16] and fs-laser writing [17]. However high-gain EDF are doped with Al or P rather than Ge that is normally responsible for high UV photosensitivity. Thus, fs writing [17] seems as the most suitable method. However, to the best of our knowledge, this technique has not been used yet to fabricate LPG in EDFs; e.g., in [17], fs-writing was used for fiber Bragg grating fabrication.

In order to observe effects described in our theoretical analysis, e.g., tuning of the mode coupling interaction strength, it is necessary to provide enough gain across the LPG length that is typically 3-20 cm long. Practically, values of at least several dB per the LPG length are required, which is somehow more than in most commercially-available EDFs.

Fortunately, there are commercially available fibers that have low numerical aperture allowing use of relatively low cost techniques such as CO_2 and arc that do not require UV-photosensitivity and have high gain per unit length (tens of dB per meter). An example is Er80-8/125 from Liekki, Finland, that has 1532 nm peak absorption of 80 dB/m and numerical aperture of 0.13.

4. EXPERIMENTAL ANALYSIS

In this subsection we show first how gain alters the transmission characteristics of an LPG in line with predictions made in Section 2.2. Following that, we compare experimental results and those obtained using the rigorous theoretical simulations discussed in Section 2.3.

4.1. Tuning of an LPG via optical gain control

This feature was demonstrated in [18] in which a 13.5 cm long LPG of 480 μm period was inscribed into a 15-cm piece of EDF (Er80-8/125 from Liekki, Finland). Fig. 5 shows obtained results for an LPG with kL=0.52π. The theoretical predictions with the gain factor being 'fitted' to get good agreement with the experiment are also shown in Fig. 5. As we see, by controlling the gain we can obtain full coupling (at the pump power of 32 mW), although the grating has strength of kL=0.52π, exactly as predicted in the theoretical analysis shown above. It is worth mentioning that as the pump current is increased from 0 mW to 32 mW, the off-resonance transmission is increased while the resonance transmission decreases. Indeed, this effect cannot be observed by simply cascading a passive LPG with an EDF-based amplifier.

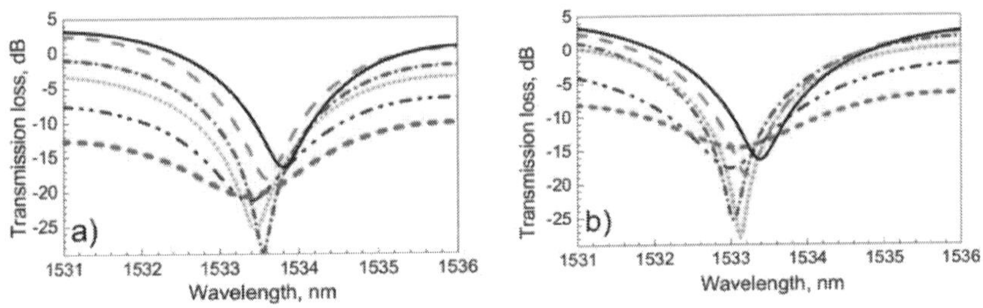

Figure 5. Measured (a) and calculated (b) transmission of LPG at kL=0.52π for input signal of 0.5 mW and different pump power levels (0 mW magenta, 8 mW blue, 17 mW green, 32 mW red, 145 mW orange, 220 mW black).

In practice, LPG-induced dip depth was varied from 7 dB (under-critical coupling) to the maximum of -28 dB (critical coupling) back to -18 dB (over-critical coupling) by varying the pump power between 0 and 220 mW.

4.2. LPG performance due to ASE and an EDF: comparison with rigorous analysis

Here we show experimental results considering an EDF theoretically modeled in the Section 2.3. Our LPG structure is shown in Fig. 6. The chosen LPG period of 480 μm corresponds to coupling into the 7th cladding mode.

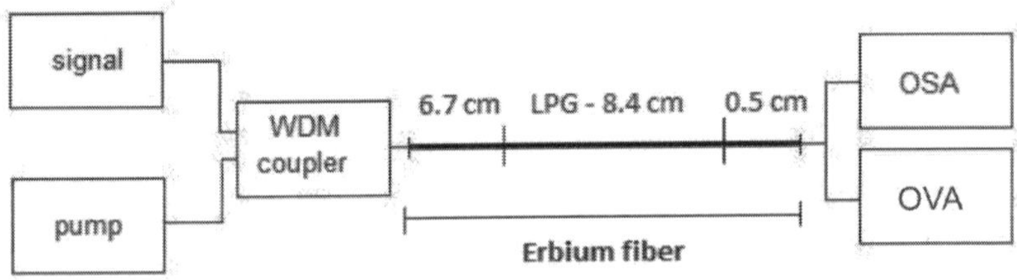

Figure 6. Experimental setup. OSA is optical spectrum analyzer; OVA is optical vector analyzer.

The theoretical and experimental results are shown in Fig. 7. For the critical coupling, there is about -33 dB of signal measured at the resonance in the core mode at the output, which is in a reasonable agreement with the theory that predicts -36 dB, Fig. 7b. These values are, however, significantly higher that those measured in a passive LPG, where <-60 dB was observed [14].

To confirm that the lower level of the transmission loss for an active-fiber based LPG as compared to a passive-fiber-based LPG is due to ASE rather than any other effect (e.g., non-uniformity of the LPG due to pump power depletion across the LPG length), we measured the phase of the transmitted light around the resonant wavelength, Fig. 8. Theory predicts that full coupling should be manifested itself as a π phase jump across the resonance. As seen in Fig. 8, a π phase-shift was observed confirming the grating is operated at the critical coupling condition in which all the light from the fiber core mode is out-coupled. This confirms that the residual light in the core mode at the resonance wavelength (which is close to -33 dB, Fig. 7) has to be due to ASE rather than due to limited coupling ability of the LPG.

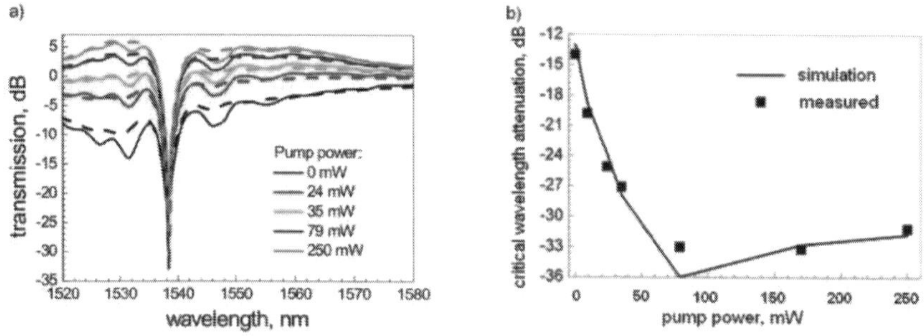

Figure 7. Left: experimental (solid) and theoretically computed (dashed) transmissions for various pump powers. Right: dependence of the maximum attainable resonant dip on the pump power (measured: dots; predicted: line)

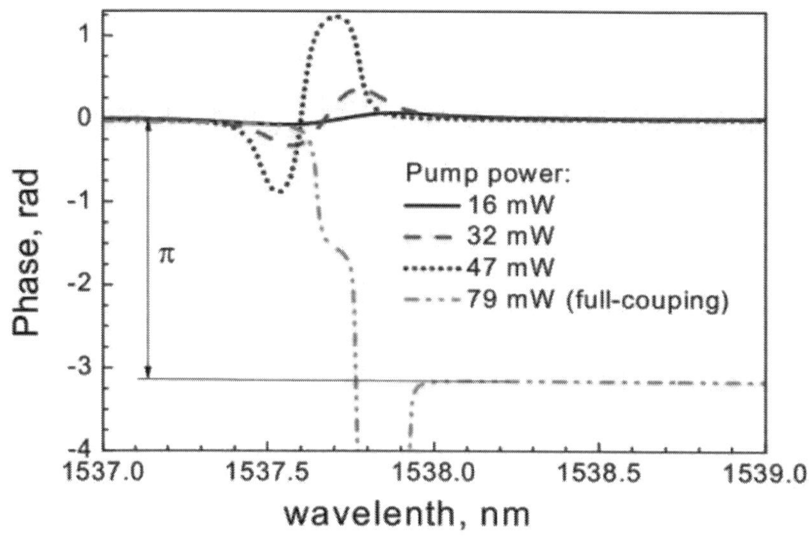

Figure 8. Phase characteristics of the fabricated LPG measured with optical vector analyzer (OVA, Fig. 6) for various pump powers.

5. APPLICATIONS

Several applications of LPGs in active fibers have been proposed. LPG directly-written into an EDF was suggested to perform flattening of the uneven EDF gain [1]. Tuning of the transmission characteristics of an LPG via optical pumping, which has been discussed in detail above, was reported in several reports (e.g.,[18-20]). This may be of interest in many applications, including all-optical signal processing, where besides active control of the grating parameters, it is advantageous to simultaneously filter and compensate for the loss that is due to the filtering process. We discuss in detail an example of an all-optical signal processing (all-optical differentiator) [20] later. Other applications are in fiber lasers, taking advantage of the fact that cladding modes have special dispersion properties that are suitable for controlling dispersion in mode-locked fiber lasers and also have large modal areas that can be exploited in high-power fiber lasers and amplifiers [21]. Another application is in using higher order modes for high-power optical amplification. In [16], the pump and the signal were simultaneously converted into a higher-order mode (a cladding mode of an inner-cladding of a dual-cladding fiber) – the LPGs for mode conversion were written in the EDF, which was doped with active ions simultaneously in the core and in the cladding.

5.1. All optical differentiation

Differentiation of the Nth order is mathematically described as $\partial^n u(t)/\partial t^n$, where $u(t)$ is the complex envelope of an arbitrary input signal spectrally centered at ω_0. The corresponding Fourier spectrum can be expressed as $(-j(\omega-\omega_0))^N U(\omega)$, where $U(\omega)$ is the spectrum of the input signal $u(t)$, ω is the optical frequency ($\omega-\omega_0$ is then the baseband). Therefore N-th order differentiation in optical domain can be obtained using a linear optical filter with a spectral transfer function proportional to $(-j(\omega-\omega_0))^N$. Unfortunately, such filter has zero transmission at the signal carrier ω_0, resulting in poor energetic efficiency (EE) defined as the ratio of the signal power after and before the differentiator. In practice, it was found that EE is <5% for an LPG-based differentiator [22]. This was confirmed by later experiments, in which some applications of this device for all-optical signal processing [23] were suggested. High order differentiation can be also realized using LPGs [24] by introducing π phase-shifts along the LPG. However, EE is getting worse as the order of the differentiation increases [24]. Obviously, LPG written in an active fiber could be very helpful in overcoming this issue.

Let us discuss advantages of LPGs directly written into active fibers on an example of optical differentiators. The use of a passive grating in conjunction with an amplifier at its output could lead to high levels of generated ASE inside the amplifier due to low EE of the differentiation process. In the case of pre-amplification of the signal prior to differentiation, emerging nonlinearities may severely impair the output signal. E.g., for pulses with FWHM < 1 ps, nonlinearities could be observed even for modest average powers. Ideal setup is thus a configuration where we perform amplification and filtration simultaneously which is the case of LPGs in active optical fibers. Indeed, the additional flexibility by tuning the LPG via controlling pump power should be also considered.

In general, the LPG structure in an active fiber is formed by a preamplifier (active fiber before the LPG), the active grating (LPG) and a post amplifier (active fiber after the LPG). Thus it is necessary to find optimal position of the grating within the active fiber to avoid the undesirable phenomena – low signal buried in the noise on one side and emerging nonlinearities on the other. In the process of designing it is necessary to take into account: total length of EDF, position of the grating within the EDF and input signal power. The total length of EDF influences the resulting gain of the output signal. Next two parameters (position of LPG and input signal power) are to be considered in combination for minimizing the ASE and avoid nonlinearities.

For theoretical analysis [4] we have chosen LPG sample with the following parameters [20]: coupling into the seventh cladding mode (the required grating period Λ=470 μm to obtain maximum coupling at 1540 nm); LPG length of 8 cm, which corresponds to the available bandwidth for differentiation to 500 GHz; input signal is Gaussian with FWHM=1.6 ps; length of EDF S=55 cm and pumping at 976 nm with power of 200 mW. In the simulation two levels of signal power were used – low (100 μW) and high (1 mW), because studied characteristics significantly differ for different levels of signal.

Energetic efficiency as a function of LPG position within EDF for two levels of signal power is shown in Fig. 9a.

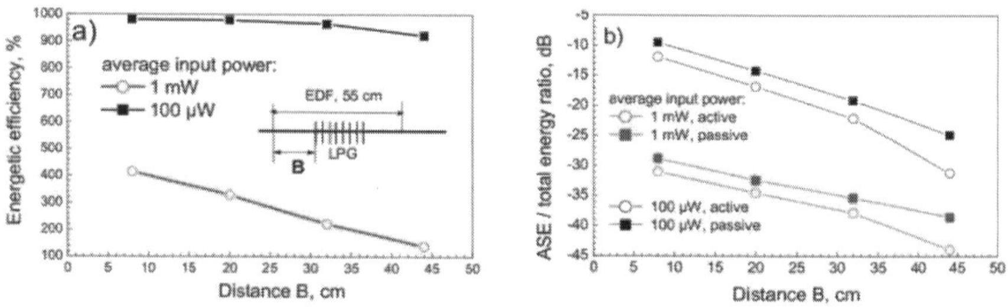

Figure 9. a) Energetic efficiency for two levels of input signal power, (b) noise characteristics for EDF with embedded passive or active LPG.

For relatively low input powers the EE is practically independent from position of the grating along the fiber. For higher signal powers the dependence on the position of LPG is clearly visible (when the grating is at the beginning or end of the fiber, EE reaches 400%and 120%, respectively). The higher EE for the grating at the beginning of EDF is compensated by worse noise characteristics as shown on Fig. 9b.

For low input power (100 µW) the ratio of ASE to the total energy (ASE+signal) at the resonance drops from -12 dB (for LPG at the beginning of EDF) to -31 dB (for LPG at the end of EDF). For higher input signal power the ASE suppression is better ranging from -31 dB (LPG at the beginning of the EDF) to -44 dB (LPG at the end of the EDF). Thus, an LPG placed at the end of EDF gets the lowest ASE, while unfortunately also having the lowest EE. This trade-off between ASE suppression and EE has to be considered depending on the requirements of a particular application.

It is interesting to compare the studied design (LPG within EDF) with a design in which a passive LPG (written in a passive fiber) would be combined directly with an pre- and post-amplifier. We consider preamplifier of length B, passive LPG with the same length of 8 cm and post-amplifier with length S-B in order to get the same amplification at the end of the structure. The input signal power is the same for both cases (passive and active LPG). Comparison, Fig. 9b, shows that the passive model has – in contrast to the active model – 2 to 4 dB worse noise characteristics.

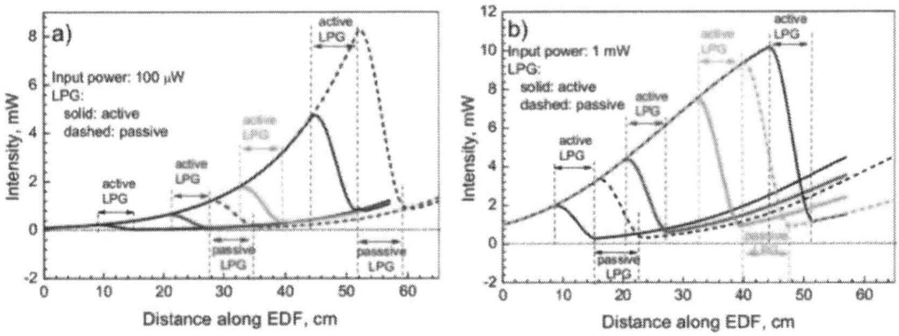

Figure 10. Power evolution for four different relative positions of the LPG along the EDF (black, red green and blue lines) for (a) 100 μW and (b) 1 mW input power. The performance considering a passive-fiber based LPG is shown as dashed lines.

For keeping nonlinearities low, it is necessary to minimize signal peak power along the EDF, which practically means keeping approximately constant level of the signal along the EDF. Fig. 10a shows the signal intensity for different position of the LPG within the EDF and the input signal power of 100 μW. At the same time it is shown that for the passive grating and the same length of EDF there are positions within the fiber with increased level of power and hence possible nonlinearities. Similarly we can observe such increased levels also for input signal power of 1 mW (Fig. 10b), although they are not as significant as in case of the weak input power. From Fig. 10 one can see that for preserving the same level of intensity along the fiber it is advantageous to place the LPG close to the middle or in the first quarter of the EDF for relatively low (100 μW) and high (1 mW) input signal power levels, respectively.

Experimental realization of a differentiator in an active LPG was reported in (Krcmarik et al., 2009). The differentiator was optimized for bandwidth of 1 THz (LPG length of 3.9 cm) and relative high input power levels (LPG within the first quarter of the EDF). The entire EDF length was 35 cm to obtain EE in excess of 100% (so-called 'loss-less differentiation'). Experimental results obtained using 0.9-ps FWHM pulses of 2 mW average power are shown in Fig. 11. For comparison, performance of an ideal differentiator is also shown there.

EE of 151% was measured. However, when the input signal was launched from the opposite side of the fiber, the EE dropped to 72%, exactly in line with the theoretical predictions discussed earlier. The processed pulse demonstrates very close fit to the theoretically predicted waveform. A sharp π-phase jump in the LPG transmission necessary for accurate differentiation was also observed.

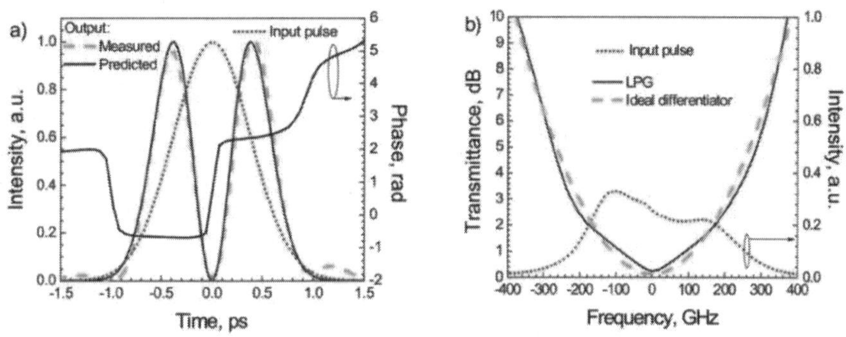

Figure 11. a) Experimental and theoretical performance in time domain [20] measured phase characteristics [25] are also shown, (b) transmittance of the LPG in the vicinity of the resonance wavelength and input pulse spectrum showing that it fits entirely into the spectral region where the LPG transmittance follows the theoretically required shape.

For evaluation of the error between the ideal and the measured waveform we used square quadratic deviation formula [26]. For forward direction the error was 2.4% (Fig. 11) and for the reversed direction the error was 2.9%.

Realizing 'loss-less' higher-order differentiators, as well as characterizing experimentally the sensitivity to non-linearities are subject to further research.

6. CONCLUSIONS

Simultaneous diffraction and gain brings many interesting new features that are still waiting to be exploited. This chapter gives a brief overview over the state-of-the art and summarizes the key properties of long period gratings made in active optical fibers.

REFERENCES

1. R. Singh, Sunanda, Sharma EK. Gain flattening by long period gratings in Erbium doped fibersOptics Communications2004123 EOF132 EOF
2. X. Daxhelet, M. Kulishov, Theory and practice of long-period gratings: when a loss becomes a gain.Optics Letters2003289686688
3. M. Kulishov, V. Grubsky, J. Schwartz, X. Daxhelet, D. V. Plant, Tunable waveguide transmission gratings based on active gain control.IEEE Journal of Quantum Electronics20041715 EOF
4. D. Krcmarík, R. Slavík, M. Karásek, M. Kuslishov, Theoretical and experimental analysis of long-period fiber gratings made directly into Er-doped active fibers. Journal of Lightwave Technology 2009
5. D Marcuse, . (1991). Theory of dielectric waveguides (2nd edition)-chapter 2, Academic Press, 0-12470-951-6 York
6. M. Kong, B. Shi, Field solution and characteristics of cladding modes of optical fibersFiber and Integrated Optics2006305 EOF321 EOF
7. T. Erdogan, Cladding-mode resonances in short- and long-period fiber grating filtersJOSA A 19971760 EOF
8. K. Thyagarajan, J. K. Anad, A novel design of an intrinsically gain-flattened erbium doped fiberOptics Communications2000407 EOF413 EOF
9. C. Jiang, W. Hu, Q. Zeng, Numerical analysis of concentration quenching model of Er3+-doped phosphate fiber amplifier. IEEE Journal of Quantum Electronics 2003
10. P. Blixt, J. Nilsson, T. Carlnas, B. Jaskorzynska, Concentration-dependent upconversion in Er3+-doped fiber amplifiers: experiments and modeling. IEEE Transaction Photonics Technology Letters 1991311996998

11. P. Myslinky, D. Nguyen, J. Chrostowski, Effects of concentration on the performance of Erbium-doped fiber amplifiers. Journal of Lightwave Technology 1997
12. E. Desurvire, Study of the complex atomic susceptibility of Erbium-doped fiber amplifiersJournal of Lightwave Technology19901517 EOF1527 EOF
13. M. Karásek, J. Ctyroký, Design considerations for Er3+-doped planar optical amplifiers in silica on silicon. Journal of Optical Communications 1995
14. R. Slavík, Extremely deep long-period fiber grating made with CO2 laser.IEEE Photonis Technology Letters 20061705 EOF1707 EOF
15. G. Rego, R. Falate, J. L. Santos, H. M. Salgado, J. L. Fabris, S. L. Semjonov, E. M. Dianov, Arc-induced long-period gratings in aluminosilicate glass fibers.Optics Letters20052065 EOF7 EOF
16. J. W. Nicholson, J. M. Fini, A. M. Desantolo, E. Monberg, DiMarcello F, Fleming J, Headley C, DiGiovanni DJ, Ghalmi S, Ramachandran S. A higher-order-mode Erbium-doped-fiber amplifier.Optics Express201017651 EOF7 EOF
17. E. Wikszak, J. Thomas, J. Burghoff, B. Ortac, J. Limpert, S. Nolte, U. Fuchs, A. Tünnermann, Erbium fiber laser based on intracore femtosecond-written fiber Bragg grating.Optics Letters20062390 EOF2 EOF
18. R. Slavík, M. Kulishov, Active control of long-period fiber-grating-based filters made in erbium-doped optical fibers.Optics Letters2007757 EOF9 EOF
19. A. Quintela, M. A. Quintela, C. Jauregui, J. M. Lopez-higuera, Optically tunable long-period fiber grating on an Er3+ fiber.IEEE Photonics Technology Letters2007732 EOF734 EOF
20. D. Krcmarík, R. Slavík, Y. Park, M. Kulishov, J. Azaña, First-order loss-less differentiators using long period gratings made in Er-doped fibers. Optics Express 2009
21. D. Sáez-rodriguez, J. L. Cruz, A. Díez, M. V. Andrés, Fiber laser with combined feedback of core and cladding modes assisted by an intracavity long-period grating.Optics Letters20111839 EOF41 EOF
22. M Kulishov, ., J Azaña, . (2005). Long-period fiber gratings as ultrafast optical differentiators. Optics Lett., 30 20 27002702 , 0146-9592
23. R Slavík, ., Y Park, ., J Azaña, . (2007). Tunable dispersion-tolerant picosecond flat-top waveform generation using an optical differentiator. Opt. Express, 15 11 67176726 , 1094-4087

24. M. Kulishov, D. Krcmarík, R. Slavík, Design of terahertz-bandwidth arbitrary-order temporal differentiators based on long-period fiber gratings. Optics Letters 2007
25. Y. Park, F. Li, J. Azaña, Characterization and optimization of optical pulse differentiation using spectral interferometry. IEEE Photonics Technology Letters 20061798 EOF
26. N. Q. Ngo, S. F. Yu, S. C. Tjin, C. H. Kam, A new theoretical basis of higher-derivative optical differentiatorsOptics Communications2003115 EOF129 EOF

Index